TEACHER GUIDE | Includes Student Worksheets | Includes: Answer Keys, Weekly Lesson Schedule, Worksheets, Quizzes & Tests

7th –8th Grade | Science

Awesome Science: Historical Geology

Master Books® Curriculum

First printing: March 2016
Third printing: September 2018

For information write:

Master Books®, P.O. Box 726, Green Forest, AR 72638

Master Books® is a division of the New Leaf Publishing Group, Inc.

ISBN: 978-0-89051-930-1
ISBN: 978-1-61458-500-8 (digital)

Unless otherwise noted, Scripture quotations are from the New King James Version of the Bible.

Printed in the United States of America

Please visit our website for other great titles:
www.masterbooks.com

For information regarding author interviews,
please contact the publicity department at (870) 438-5288.

100% GUARANTEE Faith Grower 100% GUARANTEE

Table of Contents

The Awesome Science series is brought to you by Emmy-nominated producer, **Kyle Justice**, the same producer of the *X-Nilo Show* and *The Creation Network*. His work has appeared on such networks as *National Geographic, ESPN*, and *The Outdoor Channel*. The series is narrated by Kyle's son, **Noah Justice**.

Using This Teacher Guide

Features: The suggested weekly schedule enclosed has easy-to-manage lessons that guide the reading, worksheets, and all assessments. The pages of this guide are perforated and three-hole punched so materials are easy to tear out, hand out, grade, and store. Teachers are encouraged to adjust the schedule and materials needed in order to best work within their unique educational program.

Lesson Scheduling: Students are instructed to read the pages in their book and then complete the corresponding section provided by the teacher. Assessments that may include worksheets, activities, quizzes, and tests are given at regular intervals with space to record each grade. Space is provided on the weekly schedule for assignment dates, and flexibility in scheduling is encouraged. Teachers may adapt the scheduled days per each unique student situation. As the student completes each assignment, this can be marked with an "X" in the box.

- **Approximately 30 to 45 minutes per lesson, five days a week**
- **Includes answer keys for worksheets, quizzes, and tests.**
- **Worksheets for each section**
- **Quizzes and tests are included to help reinforce learning and provide assessment opportunities.**
- **Designed for grades 7 to 8 in a one-year course to earn 1 science credit**

Course Description

Teen Noah Justice leads students across the American landscape with these 12 phenomenal DVDs, created by his father, Kyle Justice, the same producer of the "X-Nilo Show" and "The Creation Network." Students will explore the geologic marvels of the Arches and Natural Bridges, Yellowstone, the Grand Canyon and more, all showing so clearly that catastrophic processes of the Flood just a few thousand years ago have taken place to shape these amazing formations. Secular geologists continue to deny the truth of the global Flood, but geologic features around the world continue to verify that it was real. This course provides a visually stunning view of God's world and supports the authority and historical evidence of the Bible. This is a faith-affirming study that will both fascinate and educate.

Course Objectives

Students completing this course will:

- Investigate the geology of God's creation through wonders like Mount St. Helens; viewing the 12 DVD Awesome Science series that takes students all across the United States
- Review the creation-based books *Your Guide to the Grand Canyon* and *Your Guide to Yellowstone and Grand Teton National Parks* from the True North series
- Evaluate insights about paleontological sites such as John Day Fossil Beds and Dinosaur National Monument
- Explore historical studies of the native peoples of America including those at Mesa Verde and Chaco Ruins
- Demonstrate how the Rocky Mountains were formed and eroded by the catastrophic tectonics of the Flood
- Identify the largest concentrations of mammoth fossils in the world at the Mammoth Site in South Dakota
- Learn about Glacier National Park and research the evidence for this area being created and eroded quickly by the effects of the Flood

Optional Science Lab

There are a variety of companies that offer science labs that complement our courses. These items are only suggestions, not requirements, and they are not included in the daily schedule. We have tried to find materials that are free of evolutionary teaching, but please review any materials prior to presentation. The following items are available from www.HomeTrainingTools.com.

RM-GEOBAG Geology Field Trip in a Bag

RM-ROCKMIN Rocks & Minerals of the U.S. Basic Set

Map of National Parks and Monuments in the Western United States

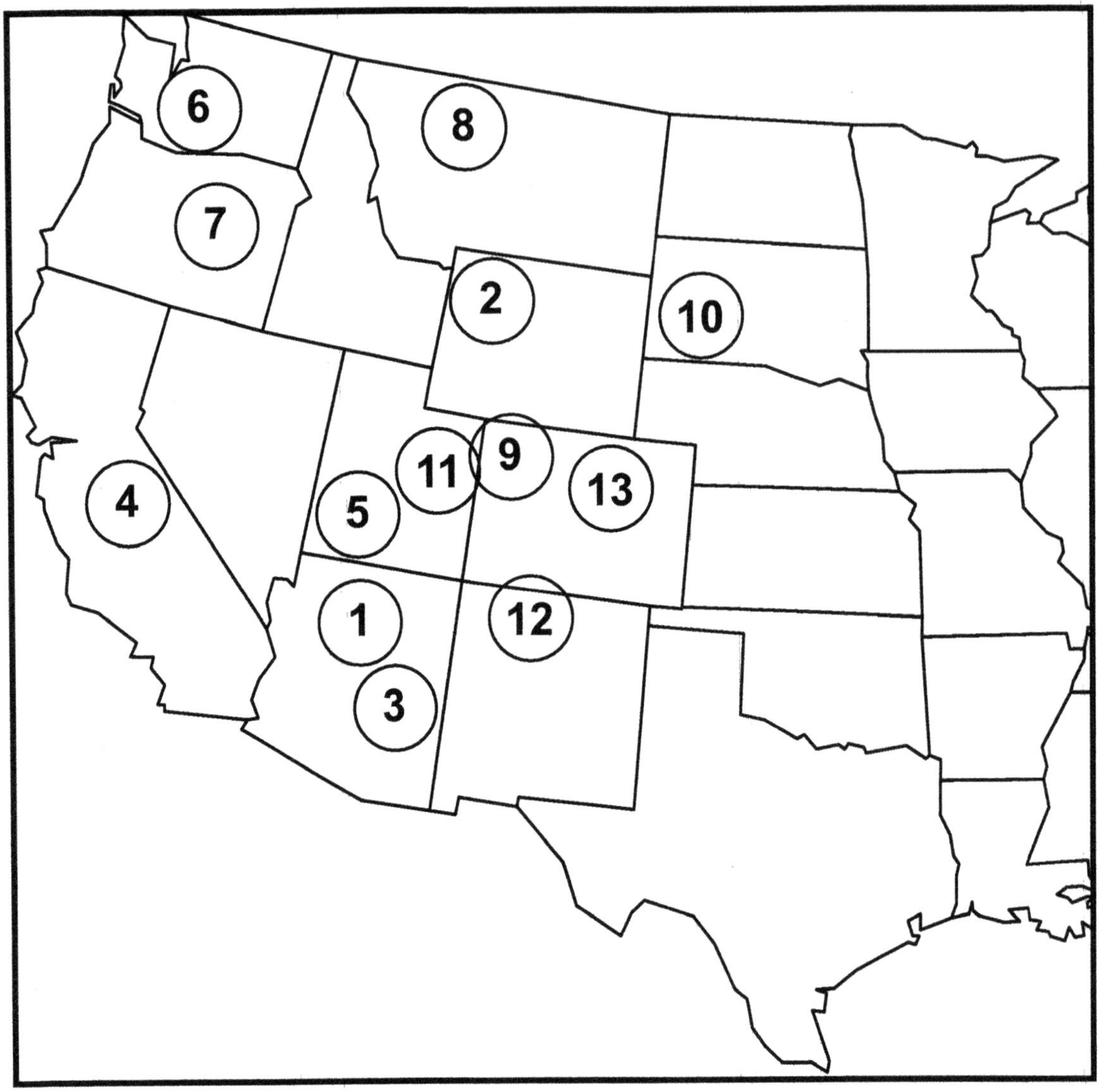

1. Grand Canyon National Park
2. Yellowstone National Park
3. Meteor Crater and Petrified Forest National Parks
4. Yosemite National Park
5. Zion National Park
6. Mount St. Helen's
7. John Day Fossil Beds National Monument
8. Glacier National Park
9. Dinosaur National Monument
10. Mammoth Site
11. Arches National Park and Natural Bridges National Monument
12. Mesa Verde and Chaco Ruins
13. Rocky Mountian National Park

First Semester Suggested Daily Schedule

Date	Day	Assignment	Due Date	✓	Grade
		First Semester-First Quarter			
Week 1	Day 1	Explore the Grand Canyon • Watch DVD (Episode 1) Complete Worksheet 1 • Pages 17-18 • *Teacher Guide* • (TG)			
	Day 2	Explore the Grand Canyon • Complete Worksheet 2 Pages 19-20 • (TG)			
	Day 3	Explore the Grand Canyon • Complete Worksheet 3 Pages 21-22 • (TG)			
	Day 4	Explore the Grand Canyon • Complete Worksheet 4 Page 23 • (TG)			
	Day 5	Explore the Grand Canyon • Complete Worksheet 5 Pages 25-26 • (TG)			
Week 2	Day 6	Explore the Grand Canyon • Complete Worksheet 6 Pages 27-28 • (TG)			
	Day 7	Explore the Grand Canyon • Complete Worksheet 7 Page 29 • (TG)			
	Day 8	Explore the Grand Canyon • Complete Worksheet 8 Pages 31-32 • (TG)			
	Day 9	Explore the Grand Canyon • Complete Worksheet 9 Pages 33-34 • (TG)			
	Day 10	Explore the Grand Canyon • Complete Worksheet 10 Pages 35-36 • (TG)			
Week 3	Day 11	Read Foreword - Section One • Pages 5-13 *True North Grand Canyon* • (TNGC)			
	Day 12	Read Section Two, Three, & Four • Pages 14 - 37 • (TNGC)			
	Day 13	Section Two, Three, & Four • Complete Worksheet 1 Pages 39-40 • (TG)			
	Day 14	Read Section Five • Pages 38 - 59 • (TNGC)			
	Day 15	Section Five • Complete Worksheet 2 Pages 41-42 • (TG)			
Week 4	Day 16	Read Section Five • Pages 60 - 79 • (TNGC)			
	Day 17	Section Five • Complete Worksheet 3 Pages 43-44 • (TG)			
	Day 18	Read Section Five • Pages 80 - 99 • (TNGC)			
	Day 19	Section Five • Complete Worksheet 4 Pages 45-46 • (TG)			
	Day 20	Read Section Five • Pages 100 - 119 • (TNGC)			
Week 5	Day 21	Section Five • Complete Worksheet 5 Pages 47-48 • (TG)			
	Day 22	Read Section Five • Pages 120 - 139 • (TNGC)			
	Day 23	Section Five • Complete Worksheet 6 Pages 49-50 • (TG)			
	Day 24	Read Section Six & Seven • Pages 140 - 161 • (TNGC)			
	Day 25	Section Six & Seven • Complete Worksheet 7 Pages 51-52 • (TG)			

Date	Day	Assignment	Due Date	✓	Grade
Week 6	Day 26	Read Section Eight • Pages 162 - 167 • (TNGC)			
	Day 27	Section Eight • Complete Worksheet 8 Pages 53-54 • (TG)			
	Day 28	Read Section Nine • Pages 168 - 175 • (TNGC)			
	Day 29	Section Nine • Complete Worksheet 9 Pages 55-56 • (TG)			
	Day 30	Read Section Ten • Pages 176 - 179 • (TNGC)			
Week 7	Day 31	Section Ten • Complete Worksheet 10 Pages 57-58 • (TG)			
	Day 32	**True North Grand Canyon – Test 1:** • Pages 291-292 • (TG)			
	Day 33	Explore Yellowstone • Watch DVD (Episode 2) Complete Worksheet 1 • Pages 61-62 • *Teacher Guide* • (TG)			
	Day 34	Explore Yellowstone • Complete Worksheet 2 Pages 63-64 • (TG)			
	Day 35	Explore Yellowstone • Complete Worksheet 3 Pages 65-66 • (TG)			
Week 8	Day 36	Explore Yellowstone • Complete Worksheet 4 Page 67 • (TG)			
	Day 37	Explore Yellowstone • Complete Worksheet 5 Page 69 • (TG)			
	Day 38	Explore Yellowstone • Complete Worksheet 6 Pages 71-72 • (TG)			
	Day 39	Explore the Yellowstone • Complete Worksheet 7 Page 73 • TG)			
	Day 40	Explore Yellowstone • Complete Worksheet 8 Page 75 • (TG)			
Week 9	Day 41	Explore Yellowstone • Complete Worksheet 9 Page 77 • (TG)			
	Day 42	Explore Yellowstone • Complete Worksheet 10 Page 79 • (TG)			
	Day 43	Explore Yellowstone • Complete Worksheet 11 Page 81 • (TG)			
	Day 44	Map of All the Sites to Study • Page 6 • (TG)			
	Day 45	**Quiz on All of the Sites** (Map) • Page 299 • (TG)			
		First Semester-Second Quarter			
Week 1	Day 46	Explore Meteor Crater & Petrified Forest • Watch DVD (Episode 3) Complete Worksheet 1 • Pages 85-86 • *Teacher Guide* • (TG)			
	Day 47	Explore Meteor Crater & Petrified Forest • Complete Worksheet 2 Pages 87-88 • (TG)			
	Day 48	Explore Meteor Crater & Petrified Forest • Complete Worksheet 3 Pages 89-90 • (TG)			
	Day 49	Explore Meteor Crater & Petrified Forest • Complete Worksheet 4 Pages 91-92 • (TG)			
	Day 50	Explore Meteor Crater & Petrified Forest • Complete Worksheet 5 Pages 93-94 • (TG)			

Date	Day	Assignment	Due Date	✓	Grade
Week 2	Day 51	Explore Meteor Crater & Petrified Forest • Complete Worksheet 6 Pages 95-96 • (TG)			
	Day 52	Explore Meteor Crater & Petrified Forest • Complete Worksheet 7 Pages 97-98 • (TG)			
	Day 53	Explore Meteor Crater & Petrified Forest • Complete Worksheet 8 Page 99 • (TG)			
	Day 54	Explore Yosemite & Zion • Watch DVD (Episode 4) Complete Worksheet 1 • Pages 103-104 • *Teacher Guide* • (TG)			
	Day 55	Explore Yosemite & Zion • Complete Worksheet 2 Pages 105-106 • (TG)			
Week 3	Day 56	Explore Yosemite & Zion • Complete Worksheet 3 Page 107 • (TG)			
	Day 57	Explore Yosemite & Zion • Complete Worksheet 4 Page 109 • (TG)			
	Day 58	Explore Yosemite & Zion • Complete Worksheet 5 Pages 111-112 • (TG)			
	Day 59	Explore Yosemite & Zion • Complete Worksheet 6 Page 113 • (TG)			
	Day 60	Explore Yosemite & Zion • Complete Worksheet 7 Page 115 • (TG)			
Week 4	Day 61	Explore Yosemite & Zion • Complete Worksheet 8 Pages 117-118 • (TG)			
	Day 62	Explore Yosemite & Zion • Complete Worksheet 9 Pages 119-120 • (TG)			
	Day 63	Explore Mount St. Helens • Watch DVD (Episode 5) Complete Worksheet 1 • Page 123 • *Teacher Guide* • (TG)			
	Day 64	Explore Mount St. Helens • Complete Worksheet 2 (Page 1) Pages125 • (TG)			
	Day 65	Explore Mount St. Helens • Complete Worksheet 2 (Page 2) Page 126 • (TG)			
Week 5	Day 66	Explore Mount St. Helens • Complete Worksheet 3 (Page 1) Page 127 • (TG)			
	Day 67	Explore Mount St. Helens • Complete Worksheet 3 (Page 2) Page 128 • (TG)			
	Day 68	Explore Mount St. Helens • Complete Worksheet 4 Pages 129-130 • (TG)			
	Day 69	Explore Mount St. Helens • Complete Worksheet 5 (Page 1) Page 131 • (TG)			
	Day 70	Explore Mount St. Helens • Complete Worksheet 5 (Page 2) Page 132 • (TG)			

Date	Day	Assignment	Due Date	✓	Grade
Week 6	Day 71	Explore Mount St. Helens • Complete Worksheet 6 Pages 133-134 • (TG)			
	Day 72	Explore Mount St. Helens • Complete Worksheet 7 Pages 135-136 • (TG)			
	Day 73	Explore Mount St. Helens • Complete Worksheet 8 Pages 137-138 • (TG)			
	Day 74	Explore Mount St. Helens • Complete Worksheet 9 Pages 139-140 • (TG)			
	Day 75	Explore John Day Fossil Beds • Watch DVD (Episode 6) Complete Worksheet 1 (Page 1) • Page 143 • (TG)			
Week 7	Day 76	Explore John Day Fossil Beds • Complete Worksheet 1 (Page 2) Page 144 • (TG)			
	Day 77	Explore John Day Fossil Beds • Complete Worksheet 2 Pages 145-146 • (TG)			
	Day 78	Explore John Day Fossil Beds • Complete Worksheet 3 Page 147 • (TG)			
	Day 79	Explore John Day Fossil Beds • Complete Worksheet 4 Fill in the Blanks • Page 149 • (TG)			
	Day 80	Explore John Day Fossil Beds • Complete Worksheet 4 True or False • Pages 149-150 • (TG)			
Week 8	Day 81	Explore John Day Fossil Beds • Complete Worksheet 5 Pages 151-152 • (TG)			
	Day 82	Explore John Day Fossil Beds • Complete Worksheet 6 (Page 1) Page 153 • (TG)			
	Day 83	Explore John Day Fossil Beds • Complete Worksheet 6 (Page 2) Pages 153-154 • (TG)			
	Day 84	Explore John Day Fossil Beds • Complete Worksheet 7 Pages 155-156 • (TG)			
	Day 85	Explore John Day Fossil Beds • Complete Worksheet 8 (Page 1) Page 157 • (TG)			
Week 9	Day 86	Explore John Day Fossil Beds • Complete Worksheet 8 (Page 2) Page 158 • (TG)			
	Day 87	Explore John Day Fossil Beds • Complete Worksheet 9 Pages 159-160 • (TG)			
	Day 88	Bonus Activity Day (complete a bonus activity you would enjoy)			
	Day 89	Review Past DVD Worksheets for First Semester Exam			
	Day 90	**First Semester Exam – Test 2** • Pages 293-294 • (TG)			
		Mid-Term Grade			

Second Semester Suggested Daily Schedule

Date	Day	Assignment	Due Date	✓	Grade
		Second Semester-Third Quarter			
Week 1	Day 91	Read Pages 6-9 • *True North Yellowstone* • (TNY) Complete Worksheet 1 • Pages 163-164 • (LP)			
	Day 92	Read Section One through Four • Pages 10 -29 • (TNY)			
	Day 93	Section One through Four • Complete Worksheet 2 Pages 165-166 • (TG)			
	Day 94	Read Section Five • Pages 30 - 49 • (TNY)			
	Day 95	Section Five, Part 1 • Complete Worksheet 3 Pages 167-168 • (TG)			
Week 2	Day 96	Read Section Five • Pages 50 - 65 • (TNY)			
	Day 97	Section Five, Part 2 • Complete Worksheet 4 Pages 169-170 • (TG)			
	Day 98	Read Section Five • Pages 66 - 73 • (TNY)			
	Day 99	Read Section Five • Pages 74 - 81 • (TNY)			
	Day 100	Section Five, Part 3 • Complete Worksheet 5 Pages 171-172 • (TG)			
Week 3	Day 101	Read Section Five • Pages 82 - 93 • (TNY)			
	Day 102	Read Section Five • Pages 94 - 101 • (TNY)			
	Day 103	Section Five, Part 4 • Complete Worksheet 6 Pages 173-174 • (TG)			
	Day 104	Read Section Six • Pages 104 - 109 • (TNY)			
	Day 105	Read Section Six • Pages 110 - 117 • (TNY)			
Week 4	Day 106	Section Six, Part 1 • Complete Worksheet 7 Pages 175-176 • (TG)			
	Day 107	Read Section Six • Pages 118 - 121 • (TNY)			
	Day 108	Read Section Six • Pages 122 - 129 • (TNY)			
	Day 109	Section Six, Part 2 • Complete Worksheet 8 Pages 177-178 • (TG)			
	Day 110	Read Section Six • Pages 130 - 137 • (TNY)			
Week 5	Day 111	Read Section Six • Pages 138 - 141 • (TNY)			
	Day 112	Section Six, Part 3 • Complete Worksheet 9 Pages 179-180 • (TG)			
	Day 113	Read Section Seven • Pages 142-145 • (TNY)			
	Day 114	Section Seven • Complete Worksheet 10 Pages 181-182 • (TG)			
	Day 115	Read Section Eight • Pages 146 - 157 • (TNY)			
Week 6	Day 116	Section Eight • Complete Worksheet 11 Pages 183-184 • (TG)			
	Day 117	Read Section Nine & Ten • Pages 158 - 171 • (TNY)			
	Day 118	Section Nine & Ten • Complete Worksheet 12 Pages 185-186 • (TG)			
	Day 119	Study Day for True North Yellowstone Exam			
	Day 120	**True North Yellowstone Exam – Test 3:** • Pages 295-296 • (TG)			

Date	Day	Assignment	Due Date	✓	Grade
Week 7	Day 121	Explore Glacier National Park • Watch DVD (Episode 7) Complete Worksheet 1 • Pages 189-190 • (TG)			
	Day 122	Explore Glacier National Park • Complete Worksheet 2 Pages 191-192 • (TG)			
	Day 123	Explore Glacier National Park • Complete Worksheet 3 Pages 193-194 • (TG)			
	Day 124	Explore Glacier National Park • Complete Worksheet 4 Page 195 • (TG)			
	Day 125	Explore Dinosaur National Monument • Watch DVD (Episode 8) • Complete Worksheet 1 • Page 199 • (TG)			
Week 8	Day 126	Explore Dinosaur National Monument Complete Worksheet 2 • Pages 201-202 • (TG)			
	Day 127	Explore Dinosaur National Monument Complete Worksheet 3 • Page 203 • (TG)			
	Day 128	Explore Dinosaur National Monument Complete Worksheet 4 • Page 205 • (TG)			
	Day 129	Explore Dinosaur National Monument Complete Worksheet 5 • Pages 207-208 • (TG)			
	Day 130	Explore Dinosaur National Monument Complete Worksheet 6 • Page 209 • (TG)			
Week 9	Day 131	Explore Mammoth Site • Watch DVD (Episode 9) Complete Worksheet 1 • Pages 213-214 • (TG)			
	Day 132	Explore Mammoth Site • Complete Worksheet 2 Pages 215-216 • (TG)			
	Day 133	Explore Mammoth Site • Complete Worksheet 3 Pages 217-218 • (TG)			
	Day 134	Explore Mammoth Site • Complete Worksheet 4 Page 219 • (TG)			
	Day 135	Explore Mammoth Site • Complete Worksheet 5 Pages 221-222 • (TG)			
		Second Semester-Fourth Quarter			
Week 1	Day 136	Explore Arches National Park and Natural Bridges National Monument • Watch DVD (Episode 10) Complete Worksheet 1 • Page 225 • (TG)			
	Day 137	Explore Arches National Park and Natural Bridges National Monument • Complete Worksheet 2 • Page 227 • (TG)			
	Day 138	Explore Arches National Park and Natural Bridges National Monument • Complete Worksheet 3 • Page 229 • (TG)			
	Day 139	Explore Arches National Park and Natural Bridges National Monument • Complete Worksheet 4 • Page 231 • (TG)			
	Day 140	Explore Arches National Park and Natural Bridges National Monument • Complete Worksheet 5 • Page 233 • (TG)			

Date	Day	Assignment	Due Date	✓	Grade
Week 2	Day 141	Explore Arches National Park and Natural Bridges National Monument • Complete Worksheet 6 • Page 235 • (TG)			
	Day 142	Explore Arches National Park and Natural Bridges National Monument • Complete Worksheet 7 • Pages 237-238 • (TG)			
	Day 143	Explore Mesa Verde and Chaco Ruins • Watch DVD (Episode 11) • Complete Worksheet 1 • Pages 241-242 • (TG)			
	Day 144	Explore Mesa Verde and Chaco Ruins Complete Worksheet 2 • Pages 243-244 • (TG)			
	Day 145	Explore Mesa Verde and Chaco Ruins Complete Worksheet 3 (Page 1) • Page 245 • (TG)			
Week 3	Day 146	Explore Mesa Verde and Chaco Ruins Complete Worksheet 3 (Page 2) • Page 246 • (TG)			
	Day 147	Explore Mesa Verde and Chaco Ruins Complete Worksheet 4 (Page 1) • Page 247 • (TG)			
	Day 148	Explore Mesa Verde and Chaco Ruins Complete Worksheet 4 (Page 2) • Page 248 • (TG)			
	Day 149	Explore Mesa Verde and Chaco Ruins Complete Worksheet 5 • Page 249 • (TG)			
	Day 150	Explore Mesa Verde and Chaco Ruins Complete Worksheet 6 (Page 1) • Page 251 • (TG)			
Week 4	Day 151	Explore Mesa Verde and Chaco Ruins Complete Worksheet 6 (Page 2) • Page 252 • (TG)			
	Day 152	Explore Mesa Verde and Chaco Ruins Complete Worksheet 7 • Pages 253-254 • (TG)			
	Day 153	Explore Mesa Verde and Chaco Ruins Complete Worksheet 8 (Page 1) • Page 255 • (TG)			
	Day 154	Explore Mesa Verde and Chaco Ruins Complete Worksheet 8 (Page 2) • Page 256 • (TG)			
	Day 155	Explore Mesa Verde and Chaco Ruins Complete Worksheet 9 • Pages 257-258 • (TG)			
Week 5	Day 156	Explore Mesa Verde and Chaco Ruins Complete Worksheet 10 • Page 259 • (TG)			
	Day 157	Explore Mesa Verde and Chaco Ruins Complete Worksheet 11 • Pages 261-262 • (TG)			
	Day 158	Explore Mesa Verde and Chaco Ruins Complete Worksheet 12 • Page 263 • (TG)			
	Day 159	Explore Mesa Verde and Chaco Ruins Complete Worksheet 13 (Page 1) • Pages 265 • (TG)			
	Day 160	Explore Mesa Verde and Chaco Ruins Complete Worksheet 13 (Page 2) • Pages 266 • (TG)			

Date	Day	Assignment	Due Date	✓	Grade
Week 6	Day 161	Explore Mesa Verde and Chaco Ruins Complete Worksheet 14 • Pages 267-268 • (TG)			
	Day 162	Explore Rocky Mountain National Park Watch DVD (Episode 12)			
	Day 163	Explore Rocky Mountain National Park Complete Worksheet 1 • Pages 271-272 • (TG)			
	Day 164	Explore Rocky Mountain National Park Complete Worksheet 2 • Pages 273-274 • (TG)			
	Day 165	Explore Rocky Mountain National Park Complete Worksheet 3 • Page 275 • (TG)			
Week 7	Day 166	Explore Rocky Mountain National Park Complete Worksheet 4 (Page 1) • Page 277 • (TG)			
	Day 167	Explore Rocky Mountain National Park Complete Worksheet 4 (Page 2) • Page 278 • (TG)			
	Day 168	Explore Rocky Mountain National Park Complete Worksheet 5 • Pages 279-280 • (TG)			
	Day 169	Explore Rocky Mountain National Park Complete Worksheet 6 (Page 1) • Page 281 • (TG)			
	Day 170	Explore Rocky Mountain National Park Complete Worksheet 6 (Page 2) • Page 282 • (TG)			
Week 8	Day 171	Explore Rocky Mountain National Park Complete Worksheet 7 • Page 283 • (TG)			
	Day 172	Explore Rocky Mountain National Park Complete Worksheet 8 (Page 1) • Page 285 • (TG)			
	Day 173	Explore Rocky Mountain National Park Complete Worksheet 8 (Page 2) • Page 286 • (TG)			
	Day 174	Explore Rocky Mountain National Park Complete Worksheet 9 • Pages 287 • (TG)			
	Day 175	Study Day for Map Quiz • Page 6 • (TG)			
Week 9	Day 176	Map Quiz of Sites • Page 301 • (TG)			
	Day 177	Study Day for Semester Test (Watch Arches again.)			
	Day 178	Study Day for Semester Test (Watch Mesa Verde again.)			
	Day 179	Study Day for Semester Test (Watch Rocky Mountain again.)			
	Day 180	**Second Semester Exam – Test 4** • Pages 297-298 • (TG)			
		Final Grade			

Grand Canyon Worksheets

for Use with

Explore the Grand Canyon with Noah Justice
(Awesome Science Series)

Chapter 2: Overview
Chapter 3: The Evidence
About the Grand Canyon

Fill in the blanks with words from the following list:

Noah's	western	reinterpret	processes	moon	present	canyon
Colorado	biblical	evidence	Genesis	scenic	evolutionary	schists
Sea	slow	worldview	Native	basement	millions	sedimentary
Arizona	expedition	secular humanism				

1. The __________ United States has some of the most amazing ________ wonders of the world: Yellowstone, Yosemite, and the Grand Canyon.

2. The Grand Canyon is found in the northern part of________________. It has a depth of more than a mile and is as wide as 18 miles. It can be seen from the ______________. It is truly an awesome sight.

3. The amazing _______________ River has its start on the western slopes of the Rocky Mountains and runs 1,450 miles to the ocean. It travels 277 miles through the canyon. Eventually it empties into the _________ of Cortez in Mexico.

4. ______________ Americans first dwelt here, but in 1540 explorers with Coronado were the first known Europeans to see the canyon. Finally in 1869 John Wesley Powell led the first ____________ in boats down the Grand Canyon in just three months.

5. The most fascinating part of the canyon is how we can see thousands of feet of ___________________ layers produced during the Flood.

6. The Colorado River also flows through the ______________ granites and __________. This has fascinated the scientific community because it's one of the few places on earth you can study what happened in the past on such a grand scale.

7. When you walk through the park, signs are up all over the place saying the canyon was formed over ________________ of years. Why?

8. Because secular scientists believe that ______________ geologic ___________ and evolution occurred over millions of years. Evolution and millions of years are parts of the religion of ___________________, in the same way that creation and thousands of years are part of ______________ Christianity.

9. __________________ geology says the _______________ is the key to the past. In other words, the processes we see today are what formed the past, over long periods of time. By doing this, they are saying there were no catastrophes in the past, like the Flood of ________________ day.

10. Here at the Grand Canyon, they say the small Colorado River was able to form this huge ___________, given enough time. Catastrophe, which means a lot of change over a little period of time (think "________________ Flood"), is something most secular scientists don't believe can happen on such a grand scale.

11. In the evolutionary ____________________________, everything developed by chance over billions of years, and creation by God is just a "myth." But some scientists, even though grudgingly, are having to ________________ their thinking, because they realize that a proper interpretation of the ____________________ just doesn't match up with the millions of years.

Discussion Questions:

1. How does a worldview reveal how one understands the geologic evidence at places like the Grand Canyon?

2. Why does the secular explanation of the small Colorado River carving the massive Grand Canyon not make sense?

3. Who were the first inhabitants of the Grand Canyon?

Bonus Activity:

See if you can learn more about the daily lives of Native Americans who lived in the Grand Canyon by looking for books at your local library or online sources. How were they affected by the terrain and climate in this area?

Chapter 3: The Evidence
Evidence: Colorado River Plausibility Questions

Please note if the following statements are true (T) or false (F).

1. _____ The Rocky River has its start in the Colorado Mountain National Park at over 10,000 feet.
2. _____ The river then winds its way down through Utah and empties into Lake Powell at about 3,700 feet.
3. _____ After leaving Lake Powell, it meanders through the Grand Canyon.
4. _____ A very large plateau rises to 780 feet to the west of the Painted Desert, called the Kaibab Plateau. The Grand Canyon goes through this.
5. _____ How does a river rise 4,000 feet to carve a canyon? A river can always run uphill. Modern-day observations can reasonably explain how a river could have done this.
6. _____ The Colorado River should have gone in a different direction.
7. _____ Some theorize that the plateau rose while the canyon was formed, but there is no evidence for this to be the case.
8. _____ In fact, most scientists are convinced the plateau rose before the canyon was cut, so some other mechanism to carve the plateau was at work.
9. _____ Then there is the water flow. When you take the volume of water of the Colorado, even in flood stage, it makes sense a river this small (compared to the size of the canyon) could have eroded away this much material over any period of time.
10. _____ Because of gravity, a river always erodes downward.
11. _____ Floods do come and change the direction of a river, but a river 18 miles wide with the current water volumes would be so shallow that it would have very little erosional power. Some major event would need to have happened to have carved this canyon.

Discussion Questions:

1. Why is it important to understand how water flow and erosion can create land features in trying to see if the Colorado River could have carved the Grand Canyon?

2. What is gravity, and how does it impact how a river flows?

3. What is the process known as erosion?

Bonus Activity:

See if you can find images of the Painted Desert. Why was it given this name, and how is this coloration made possible?

Chapter 4: Biblical Flood
The Biblical Record

The Bible tells us that about 4,500 years ago, the world was an awful and ungodly place, leading God to be grieved because He made man. God was grieved because of man's rebellion and that he was bent toward violence, wickedness, and evil. The Bible (Genesis 6) highlights the following points:

- "Then the LORD saw that the wickedness of man was great in the earth, and that every imagination of the thoughts of his heart was only evil continually."
- "And it repented the LORD that he had made man on the earth, and it grieved him at his heart."
- "And the LORD said, I will destroy man whom I have created from the face of the earth; both man, and beast, and creeping thing, and the fowls of the air; for it repenteth me that I have made them."
- God was going to judge the world.
- God would send a global Flood to destroy mankind, except for one man and his family.
- Noah found favor with God because he was a righteous man.

Bonus Activity:

Read the following biblical passages

Genesis 6:8–12 Genesis 6:18–20
Genesis 7:2–3 Romans 12:19

Discussion Questions:

1. Originally, it was just Adam and Eve who sinned in the Garden. Yet by the time of Noah's Flood, the Bible tells us that every intent in the heart of man was evil. What does this tell us about sin?

2. God was preparing to judge the world because of sin. God is perfect and without sin — and as long as mankind was sinful, they could have no relationship with God. How does this tie back into the events around the Garden of Eden and the original Fall of man?

3. Why were Noah and his family to be saved from the coming Flood?

Chapter 5: The Layers
Chapter 6: Receding Flood
Looking at the Evidence — the Canyon Walls

Fill in the blanks with words from the following list:

recessional	creation	crust	scavengers	pressure	volcanic	water
ocean	below	basement	layers	recede	sedimentary	sea level
canyon	sediments	megascopic	sand	metamorphic	decomposition	

1. The key to help us understand what carved the canyon may be in the ____________ of the canyon walls and the ________________ rocks at the bottom.

2. There are up to 40 major ________________________ layers in the Grand Canyon area.

3. Sedimentary layers are rock layers formed by ___________ laying down silt, mud, and ________. The layers were soft at one time.

4. Once the water departed, the ___________________ hardened into rock layers, though some lower layers surely began forming into rock prior to being dry due to the ________________ from above them.

5. ________________ the sedimentary layers at the Grand Canyon are the basement rocks of granite and schist. The basement rocks are not sedimentary, but often ________________ or ________________. They were there before the sedimentary layers were laid down. They usually include granites.

6. What's most interesting about the basement rocks is that they don't contain any ________________ marine fossils, meaning they were not formed during the Flood, but were part of the original ________________, about 1,600 years earlier.

7. Signs at the Grand Canyon will tell you that the sedimentary layers seen in the ______________ walls were laid down on the basement rocks over millions of years by ________________.

8. Oceans would come in, then ________________, leaving layers, and repeat the same process at least 7 times, one on top of each other, over and over.

9. But this has problems since ocean floors are not places known to form fossils because of ________________ and ____________________________, and the Grand Canyon layers do have fossils.

10. Since oceans do not rise above the ground now 4,000 to 7,800 feet above ______________, secular scientists say that the ground was lower, then pushed up after the oceans left their deposits.

11. As creationists, we agree in one sense with secular scientists; the earth's ____________________ has risen and fallen. But we believe this all happened during the Flood and especially during the ______________________________ stage, as mountains rose and valleys sank.

Using the numbers 1 to 14, put the sequence of the great Flood in order.

_____ At the end of those 150 days, the waters began to recede, which took about five months. It took another two months of drying time for the land to appear.

_____ The fountains of the great deep burst forth, which included water and volcanic activity.

_____ The ark rose above on the waters, and the floodwaters were sustained for 150 days.

_____ Then the volume of water became less and began to channelize, forming canyons and some valleys.

_____ Once they were on the ark, the Flood came seven days later.

_____ God told Noah to build an ark.

_____ Giant water currents carried silt and mud across the globe.

_____ Sea creatures were rapidly buried by sediment and fossilized.

_____ As the catastrophe began to wind down, the moving continents came to a stop, buckling rock layers and pushing up mountains thousands of feet in just days.

_____ While the water covered the entire earth, continents moved under the oceans.

_____ Noah was to fill the ark with two of every land-dwelling, air-breathing animal (including dinosaurs) and seven of every bird and clean animal.

_____ Water ran off the continents, causing huge sheet erosion, pulling silt and soft soil into the oceans.

_____ Water became trapped in valleys between the mountains and formed gigantic inland lakes.

_____ It rained for 40 days and nights.

Chapter 5: The Layers
It's the Timing!

Fill in the blanks with words from the following list:

marine	boundaries	cross-beds	time	desert	quickly	bent
burrowing	pebbles	wet	sequence	fast	sandstone	geologic
uplifting	sinking	root systems	fossils	bottom	deterioration	

1. We do see some ________________ layers, but not large-scale fracturing or snapping, indicating the layers were moved and bent while still _________, prior to being fully hardened into rock.

2. The whole ________________ of layers had to be deposited rapidly, and then bent immediately before hardening occurred.

3. During the recession stage of the Flood, many areas of the earth's surface were going through great ______________ or ______________, causing some of this bending of the soft layers. Many mountains were formed, for example, by day 150 of the Flood.

4. When we see evidence of ______________ fossils in these sedimentary layers, it helps us realize that the ocean waters indeed covered this region quickly in order to bury these creatures, so they fossilized.

5. ____________________ are created when an animal or vegetation is buried quickly in sediment. If not quickly, then scavengers will eat the remains or they will rot and decay, but NOT be fossilized.

6. But we don't see large-scale erosion at the __________________ of many layers at the Grand Canyon. Most layer surfaces are very even, one on top of the other, with hardly any erosion __________________. It's like they were laid down very quickly on top of each other, with no time for erosional features to develop.

7. Furthermore, there should be evidence of animals ____________________ into these layers.

8. In addition, vegetation would have built up ________________________.

9. But such features are absent in the Grand Canyon. The layers are simply stacked one on the other without any evidence of large ______________ breaks between them.

10. Another challenge, in some sedimentary layers, especially near the______________ of the canyon, is that there are many large embedded boulders in them, some of which are 15 feet in diameter. This would mean there was fast-moving water carrying these large stones along.

11. The Shinarump Conglomerate covers 100,000 square miles and is full of______________. So the water that formed these layers was likely moving extremely ____________ over a very wide area in order to round so many pebbles by erosion.

12. The sedimentary layers were laid down _____________, one on top of the other.

13. Many Grand Canyon sandstones contain _________________ which are sedimentary layers that are laid down as tilted or inclined layers during their deposition, and are usually caused by water or wind. This widespread _______________ feature is often seen in ripples, dunes, anti-dunes, sand waves, bars, and delta slopes.

14. Some of these_________________ layers at the Grand Canyon are thought by secular scientists to have formed by wind in arid _______________ with blowing sand.

Bonus Activity:

Read Psalm 104:8–9 and Genesis 7:19–20 and 8:4. How do these verses fit in with the biblical account of the Flood and the formation of places like the Grand Canyon?

See if you can find pictures of erosion, dunes, and the Shinarump Conglomerate.

Chapter 5: The Layers
Dunes

Please note if the following statements are true (T) or false (F).

1. _____ We know from empirical scientific testing that windblown sand dunes achieve an angle of at least 33–34 degrees on the front faces of the dunes.
2. _____ All of the cross-beds in the Grand Canyon sandstones are at the steep angle produced in desert dunes.
3. _____ Underwater sand waves have front faces with angles less than 10 degrees.
4. _____ This is exactly what we find at the Grand Canyon. This sandstone layer was created underwater.
5. _____ These layers at the Grand Canyon are a part of what's called "The Colorful Staircase."
6. _____ There are 10–12,000 feet of sedimentary layers in the Grand Staircase, starting at the bottom of the Grand Canyon, and the top layer is at Bryce Canyon in Utah.
7. _____ All layers appear to have been laid down quickly due to the features we have already discussed.
8. _____ Secular scientists use the fossil record to tell the story of evolution, but the same evidence is better explained by quick burial of animals during the Flood.
9. _____ In the fossil record we find buried animals and plants.
10. _____ This is not what we would expect to find in light of the biblical account. First, shallow marine organisms, then fish, which were asphyxiated by a lack of oxygen, then animal tracks, and finally land animals in the bottom layers.
11. _____ Normally when an animal dies, its remains decompose and are eaten by scavengers. Over millions of years there should be no remains.
12. _____ But if decomposition and scavengers were absent due to rapid burial, a fossil could then easily have been made.
13. _____ At the bottom of the canyon is an area called the "Great Conformity." The layers below the horizontal sedimentary layers are inclined, as though a great upheaval had happened.
14. _____ Then these layers were eroded off flat with little time to be subject to further erosion of valleys or other features. The muds, silts, and sand were then quickly laid on top.
15. _____ At the beginning of the Flood, we think that the earth experienced a huge upheaval. The fountains of the great deep burst forth. The land shifted, but none of the continents moved. This was just a local event.
16. _____ The sedimentary layers of the Tapeats Sandstone and Redwall Limestone stretch across the continents of North America, Europe, and parts of Asia.
17. _____ Traces of the Coconino Sandstone can be traced to Oklahoma, and some sand grains as far as the Northern Appalachians.
18. _____ Secular scientists say an ancient river carried the sand 1,800 miles, but there is no evidence, so that must mean those scientists are right!

19. _____ A global Flood could have carried this sand west. Current direction indicators in many strata have been interpreted to show the waters were moving from the NE, directly from the northeastern part of America.

20. _____ The Bible best explains so many of the geologic features we find in the geologic column, including how these vast sandstones were formed.

Science, it's awesome!

Discussion Questions:

1. Do secular scientists and creation scientists look at different evidence because they explain how the canyon formed differently?

2. Why do these scientists not agree on one explanation for how the Grand Canyon formed if everyone looks at the same evidence?

3. Why is it important for people to know the Bible matches up with the evidence we see around us when it comes to understanding our planet's history?

Chapter 6: Receding Flood
The Two Lakes

Fill in the blanks with words from the following list:

weather	weakened	post-Flood	leaches	rainfall	polar	Kaibab
miles	greater	limestone	soft	hurricanes	stabilizing	Canyonlands
saturated	evidence	Grand Canyon				

1. It is now believed that two large lakes formed behind the uplifted __________ Plateau.

2. _____________ has been left behind of two main lakes, holding an estimated 3,000 cubic miles of water, three times that of Lake Michigan. In addition to the lakes, there's strong evidence that we interpret as harsh ______________ right after the Flood. Most creationists believe there was a ________________ Ice Age.

3. To aid this, there may have been supersized ________________ that roamed the earth as weather patterns were _____________, causing massive flooding and producing enough precipitation from the warmer climate in the early post-Flood world to create the _______________ ice sheets in a few hundred years.

4. The __________________ in the warmer regions would build these lakes to overfilling.

5. Since the plateau blocking the lakes had _______________ in it, and water ______________ limestone, the integrity of the plateau began to ______________, probably over several years. The limestone could have also still been _________________ right after the Flood, making it even more susceptible to quick erosion.

6. At some point, the integrity of the Kaibab Plateau gave way and the ________________ began to be cut.

7. As when most dams break, once the first waters were through, all the water behind it pushed forward, causing even a ________________ amount of erosion. It is thought that Hopi Lake broke through first, then _______________________ Lake. In just a few days, over 1,000 cubic _______________ of earth was eroded away.

8. Since this massive erosion was not long after the Flood, many of the layers in the Grand Canyon were still saturated and comparatively ______________, so erosion could have been very quick.

Bonus Activity:

Take sand and dirt, and place it in layers in a plastic pan. Take your hand and shape the top into an uneven surface. Be sure to add in some small and larger rocks in the mix of dirt. On one end of the pan, use your hand to make a small circular hole like a small lake with a small edge along the top. Then take a pitcher of water and slowly fill up the "lake." Get the water right to the top of the lake, and stop for a moment. Notice how the surface of the "land" in the pan appears. Then quickly pour the remaining water in the pitcher in the "lake." Did part of the rim get washed away when it overflowed? How did the surface of the "land" change after the lake overflowed?

Chapter 7: Mount St. Helens
Examples at Mount St. Helens

We can look at other examples of large canyons to help answer some of the questions about how it may have been formed. For example, was the Grand Canyon formed over millions of years by the small Colorado River or during an event like Mount St. Helen's mini grand canyon, only on a larger scale? Let's discuss it with the following questions!

1. Empirical scientific methodology means you can observe it and should be able to repeat something over and over. Even though the formation of the Grand Canyon was not observable and we cannot repeat it, we've seen other canyons on a smaller scale form before our eyes. This is what we find at Mount St. Helens in southern Washington State. Why are the events surrounding the eruption of Mount St. Helens so important versus the rock layers or other evidence?

2. Mount St. Helens was a dormant volcano. In 1980 it quickly came to life, and a major eruption occurred, blowing out the north side of the mountain and creating a blast zone unlike anything seen in recent history. What is the difference in a dormant and an active volcano? How important do you feel it was for scientists to have an opportunity to view the processes in real time that they feel shaped the earth, especially those who have a biblical worldview about it?

3. Through a giant landslide and many eruptions, several hundred feet of pulverized rock and ash were deposited. It became a wasteland. How does this barren landscape remind us of what might have happened right after the great Flood receded?

4. After the main eruption in 1980, Mount St. Helens went quiet for about two years. During this time, snow built up within the crater. The crater rim was like a large earthen dam that held back the snow and water for two years. Compare this to the "lake" you created earlier in this study guide. Why was it almost a certainty that the earthen dam at Mount St. Helens would be destroyed? (Hints: The water being collected and the volcanic activity at the site.)

5. Then in 1982 the mountain became alive again, melting the snow and ice. The water quickly eroded through the soft volcanic soil and came cascading across the blast zone. It quickly created canyons in the volcanic rock and ash layers. See if you can find any video footage of the eruptions of 1980 and 1982. Notice the shape of the mountain before the 1980 eruption and the surrounding landscape — how did these eruptions impact the land and wildlife in the area?

6. Even though the Grand Canyon has sedimentary layers, the same principles of erosion apply — lots of water cutting deeply in a little bit of time. Discuss what you think based on what you have learned. Little water and a long time? Or a lot of water and a short amount of time? What makes the most sense to you?

Chapter 8: Cavitation
Examples at Glenn Canyon Dam and Cavitation

A fact of nature that helps us better understand how water could cut through solid rock very quickly is a process called "cavitation."

In the spring of 1983 the Colorado River went into flood stage, filling Lake Powell at 148,000 cubic feet of water per second. Glenn Canyon Dam was in danger of overflowing. The power plant was running at full capacity, moving 28,000 cubic feet of water per second through the turbines.

Engineers opened the four outlet tubes, which increased the flow by 17,000 cubic feet of water per second, but much more was needed to prevent disaster. The 40-foot diameter tunnel spillway was opened and the flow was eventually increased to 32,000 cubic feet of water per second.

Scientists were aware of cavitation, a dangerous process where water can tear apart almost any surface or material when enough pressure and flow is taking place.

What Happened Next?

Engineers were worried this much flow would destroy the spillway tunnel of Glenn Canyon Dam if opened too much. Their worst fears were realized when the outgoing water became red and the ground started to shake. They saw pieces of bedrock and concrete being hurled into the air at the bottom of the tunnel. This all happened in just a few minutes. They quickly shut down the spillway and went in to investigate. What they found totally surprised them.

1. The tunnel, constructed of three-foot-thick steel-reinforced concrete, had been penetrated by pits. The sides of the tunnel were missing, and the re-bar was torn into pieces. Take a ruler and measure three feet. Think about the sides of the tunnel being that thick and made of concrete — what does that make you think about the power of water being able to rearrange the landscape?

2. Where the tunnel elbow was, near the bottom, there was a giant hole 32 feet deep, 40 feet wide, and 150 feet long. Estimates were that cavitation destroyed the tunnel at 1,000 cubic feet per second during the peak. Does this change or support what you think about a little water over a long time or a lot of water in a short time?

3. When water comes down a surface at high pressure and flow, if it hits an uneven surface, the water is thrown upward by the obstruction, which causes a partial vacuum, and a vapor cavity develops, which starts to wear away at the surface. Once a pit is made, it causes more upward waves, which makes more pits, and there begins a leapfrog process of erosion. Having read this explanation, can you explain why cavitation is just a destructive process?

4. The tunnel had to be quickly repaired. It took 63,000 cubic feet of concrete to fill. This shows us that given enough water flow and pressure, cavitation can create huge erosional features in a very short amount of time. What about other destructive water events — like floods, or tsunamis, or large waves? Can they impact the landscape as well? What do you think?

5. The water flow at Glen Canyon Dam was nothing compared to the amount of water behind the two lakes spilling across the Kaibab Plateau to create the Grand Canyon. The process of cavitation would have torn the ground apart beyond imagination. Look at a picture of the Grand Canyon. See what information you can find about how tall the cliffs are. Answer the following essay question: What geologic processes helped form the Grand Canyon, and why?

Chapter 9: Mars Canyon
Example of Mars

Please note if the following statements are true (T) or false (F).

1. _____ There are similarities in canyons on the planet Mercury like those of the Grand Canyon.
2. _____ Surprisingly, on Mars we see what looks like huge canyons formed by water, but there is very little water found on Mars, and when there is some, it's pretty much just ice.
3. _____ Ninety percent of the earth's surface is still covered by water.
4. _____ Secular scientists deny that a global Flood happened in the past as presented in the Bible.
5. _____ The Bible tells us why these scientists ignore the truth. We read in 2 Peter 3:4–6: "They will say, 'Where is this "coming" he promised? Ever since our fathers died, everything goes on as it has since the beginning of creation.' But they deliberately forget that long ago by God's word the heavens existed and the earth was formed out of water and by water. By these waters also the world of that time was deluged and destroyed" (NIV).
6. _____ If they can deny the Flood and God's judgment by the great Flood in the past, they can then deny the biblical record . . . and most importantly they think they can deny the Second Coming of Jesus Christ and the second judgment that's coming to this earth.

But we've shown that the Bible can be trusted, so Jesus is coming back! Come quickly, Lord Jesus!

Bonus Activity:

See if you can find a NASA image of Mars, the red planet. If you can find an image of Mar's surface that is even better — so you can see for yourself the canyons that exist there and the channels that show water once flowed on the planet. Find any library books you can to learn more about the surface of Mars.

Are there any other planets in our solar system that show evidence of flowing water or canyons that were formed by mega-floods?

Discussion Question:

Why do secular scientists choose to believe so quickly that the canyons on Mars were possibly formed by mega-floods without the huge amount of water to be found, yet they don't think the same sort of event caused the formation of the Grand Canyon?

Grand Canyon Worksheets

for Use with

Your Guide to the Grand Canyon (True North Series)

Terms to Know (Write the definition in the space below.)

schist

granite

unconformity

Fill in the Blank

1. If a global flood actually occurred, we would expect to find billions of _______ ________ buried in rock layers laid down by _____________ all over the _____________.

2. The Bible says that we are stewards of the earth. When visiting any public or national park a good rule of thumb to follow is "take only _______________ and leave only _______________."

3. Mules are unique animals. They are a hybrid between a _______________ and a _____________.

4. Mules are an excellent mode of transportation when visiting the Grand Canyon because they can outwork a horse by three times and their __________________ is amazing!

5. First Thessalonians 5:21 says, "But examine everything __________________, hold fast to that which is ___________" (NASB).

Discussion Question

1. Why are secular scientists still debating the merits and problems of different theories in an attempt to solve the riddle of the origin of the Grand Canyon?

Activity

1. Review the Scripture passages on page 19 of the book. Choose your favorite and create a poster or piece of art with that verse as your theme.

Terms to Know (Write the definition in the space below.)

land bridge

Tapeats Sandstone

Great Unconformity

Fill in the Blank

1. The Colorado River is ________________ feet below Mather Point.

2. The Grand Canyon is up to a mile deep, _________ miles longs, and as much as _______ miles wide.

3. As you look at the Canyon, you see 4,000 vertical feet of flat, horizontal layers. These layers rest upon igneous and metamorphic rocks called ______________________ _____________.

4. Condors have a phenomenal wingspan of ____________ feet!

5. What is the "glue" of the desert?

Discussion Question

1. Discuss why biocrusts are often called the "glue of the desert."

Activity

1. Watch sedimentary layers form before your eyes!

   Materials needed: Clear jar or wide-mouth bottle, small rocks, sand, dirt, gravel, water, ruler, watch or timer. It is recommended you do this outside in case of spills.

   Fill a jar one-third full with the rocks, gravel, dirt, and sand. Add water and stir thoroughly. Set the jar in a safe place where it will not be disturbed. Begin your timer. How long does it take for the layers to form?

Terms to Know (Write the definition in the space below.)

biological mutualism

granite gorge

nautiloid

Short Answer Questions

1. One of the best-adapted animals to life in the desert is the __________________ _________.

2. Who led the first expedition that ran the Colorado River?

3. The Clark's nutcracker has what specially designed feature?

4. Both the Havasupai and Hualapai, tribes native to the Grand Canyon, have retained legends of a ____________ ____________.

Discussion Questions

1. There are billions of nautiloid fossils ranging in size up to five feet in length and only found in one 7-foot thick layer near the base of the Redwall Limestone. Evolutionary geologists have assumed that the Redwall Limestone represents slow deposition over millions of years. Discuss how the nature of these fossils suggests that a catastrophic event was instead the cause.

2. There are many legends of a global flood. Why do you think this is significant?

Activities

1. Birds of prey are also known as raptors. Using independent research, create a poster or slideshow identifying at least four characteristics of a raptor. Be sure to include pictures!
2. Create a diagram of rock layers including the acrostic from page 79 of the book.

Terms to Know (Write the definition in the space below.)

cross-beds

Tower of Babel

butte

mesa

gorge

Short Answer Questions

1. What is the preferred diet of the rock squirrel?

2. What is the only insect known to make north-south annual migration, as do birds?

3. What kind of rock makes up the red cliffs seen throughout the Canyon?

4. Some desert plants, like the ______________ ______________, can live as seeds for up to 20 years.

5. Archaeologists have determined the Tusayan Village was inhabited around 1,000 years ago by what people?

Discussion Question

1. Read Genesis 9:1–7 and Genesis 11:1–9. How were Noah's descendents rebelling against God by building the tower of Babel? What was the result of their rebellion? How can we apply this to our lives today?

Activity

1. Research the Monarch Butterfly! Can you explain their life cycle? Why do they migrate? Where do they go? How far do they travel, and what do they do once they get there? Once you have the information find, a creative way to present it to your family. You could create a slideshow, poster, or poem, or make your own board game or crossword puzzle using facts about the Monarch as clues. Use your imagination!

Terms to Know (Write the definition in the space below.)

sheet erosion

water gap

sinkhole

Short Answer Questions

1. How can the basalts at the bottom of the Canyon be younger than the basalts at the top?

2. Coyotes make their greatest contribution to the ecology of the Grand Canyon by eating ____________________.

3. What is the significance of Cedar Mountain?

4. Which birds, considered one of the most intelligent birds alive, are often seen soaring along the rim of the Canyon?

5. What is the highest point in Arizona?

Discussion Questions

1. Discuss how the mystery of the meadows on the North Rim of the park can be explained in thousands of years and not millions.

1. Discuss how sinkholes are formed in the Grand Canyon plateaus.

Activity

1. Ancient Alphabet Hangman. Early Native Americans had their own way of writing that consisted mainly of symbols. See if you can create your own alphabet using symbols! Play Hangman with your new alphabet.

Terms to Know (Write the definition in the space below.)

symbiotic relationship

folds/folding

stream

Short Answer Questions

1. Why did the Anasazi disappear from the Grand Canyon?

2. The Anasazi Indians lived on the ________ during the summer and on the ______________ ____________ in the winter.

3. The Manzanita shrub has the ability to move its leaves throughout the day to keep its leaf edge perpendicular to the sun, thus reducing ____________________.

4. When was the Grand Canyon declared a national monument?

5. After a fire, what is the source for new growth of the area?

Discussion Questions

1. How is the top of the Dox Sandstone evidence for a global flood?

2. Explain how the folds like the one in Carbon Creek contradict the millions of years theory of the age of the earth.

Activity

1. Anasazi Art! The Anasazi are known for their beautiful basket making and pottery. Create your own "Anasazi masterpiece" by using the designs on page 131 as inspiration.

   Materials needed: Plain terra cotta pot or plain finely woven basket, acrylic paints, paintbrushes. (Be sure to cover your work surface with newspapers before beginning.)

Terms to Know (Write the definition in the space below.)

sapping

weathering

circular reasoning

Short Answer Questions

1. Name two other geological sites that provide examples of how the catastrophic processes that formed the Grand Canyon have shaped similar landscapes.

2. True scientific knowledge advances by careful observation and testing that can be repeated by others, a process that can be referred to as ____________________ __________________.

3. The sediments that form the rock layers in the Grand Canyon were transported by moving water, deposited, and finally "cemented" together by _______________ ____________________, ________________, or ____________ ____________________.

4. The difference between a contact and an unconformity is _____________.

5. How would a rapid flow of water during a catastrophic process explain the lack of rockfall at the bottom of the Canyon?

Discussion Questions

1. Have a discussion comparing and contrasting the two creationist hypotheses regarding "a lot of water over a short period of time."

2. Compare and contrast the evolutionist and the creationist theories for the origin of the layers seen in the Grand Canyon.

Activity

1. Create your own diagram of the rock layers of the Grand Canyon. Use the image on page 160 of the book as a guide.

Terms to Know (Write the definition in the space below.)

phyla

kind

Short Answer Questions

1. The sudden appearance of most every "kind" of animal found today, with no sign of any transitional predecessors, is call the "________________ ________________" by the evolutionist.

2. Charles Darwin believed the lack of ________________ ____________ was the most serious and obvious objection that could be brought against his theory.

3. Instead of "proving" evolution, the fossil record actually supports the ______________ ___________.

4. Billions of fossilized ______________ are entombed in a seven-foot-thick layer near the base of the massive Redwall Limestone formation.

Discussion Questions

1. Discuss how the orientation of clam and coral fossils found in the Grand Canyon give evidence for a catastrophic flood.

2. Discuss how the fossilized nautiloids in the Redwall Limestone give evidence for a catastrophic flood.

Activity

1. Research fossil sites located around the world. What is the nearest fossil site to your location? Perhaps you can plan a field trip! What kind of fossils are found there? Are there other geological evidences of a global flood in your location?

Terms to Know (Write the definition in the space below.)

Riparian

biological province

region

ecosystem

Short Answer Questions

1. ____________________ has always been the biggest factor controlling the ecology of the Canyon.

2. What three deserts meet at the Grand Canyon?

3. The ______________ and _________________________ family of plants have a dual-root system that includes fine roots under the surface and a deep tap root. This gives them an amazing capacity to expand to take on water immediately after a rain and also to store it for later use.

4. The tassel-eared squirrel lives mainly on ponderosa pines, using them for both ____________ and _________________.

5. What bird has been specifically designed to swim on the water or in the water and to walk along the bottom of the Colorado River?

Discussion Question

1. In what ways has the construction of Glen Canyon Dam affected the ecology of the Grand Canyon?

Activity

1. Create your own desert ecosystem by making a succulent dish garden! Research plant compatibility and preferred soil and growing conditions. How much sun and water do they need to thrive? You might be surprised! You can inexpensively purchase the needed materials at most garden centers or home stores. A great resource for information is www.cactiguide.com. Not only will your dish garden be beautiful, but it will be educational as well!

Short Answer Questions (Write the definition in the space below.)

1. The first evidence of man's use of the Canyon is by the basket-making people known as the ____________ ________________.

2. The first people to actually inhabit the Canyon were the ________________ ________________, more commonly known as the ________________.

3. John Wesley Powell made his first historic trip through the Canyon in ______________.

4. Which president designated the park a national monument in 1908?

Discussion Question

1. In viewing the Grand Canyon, we can easily observe a small part of what lies beneath the earth's crust. We can also see the tremendous power unleashed by the Flood, which carved a vast canyon through the landscape. How should a Christian respond to these observations? (See Ps. 96:3–6, Rev. 4:11, and Rom. 11:33–36.)

Activity

1. Research and plan a mock vacation to the Grand Canyon. How many miles is it from where you live? How long would it take to drive there? Would you camp out or stay in hotels? How much would food and fuel cost? Plan an itinerary and budget for the trip. Who knows, maybe one day you will get to actually take your "mock" vacation!

Yellowstone Worksheets

for Use with

Explore Yellowstone with Noah Justice
(Awesome Science Series)

Chapter 2: Overview
About Yellowstone National Park

Fill in the blanks with words from the following list:

geysers	sedimentary	eruptions	caldera	geologic	mud pots	plates
sin-cursed	basement	Flood	lava chamber	tuff	volcanic rhyolite	
super volcanoes						

1. When thinking of Yellowstone National Park, most people usually remember the ___________ and ___________, but there's so much more to see. It covers over two million acres, and three states cross its boundaries.

2. Yellowstone contains one of the world's largest ______________. The last of three gigantic __________ was hundreds of times greater than that of Mount St. Helens, and the first was thought to have been 2,000 times that of Mount St. Helens.

3. The Bible tells us that at the beginning of the worldwide Flood the fountains of the great deep burst forth. During the Flood and right after, the earth was going through tremendous ___________ changes, because of the moving of earth's outer ___________.

4. Volcanoes were still active, shaping the land after the ______________. We can see other super volcanoes around the world. They can be seen on almost every continent.

5. Besides the ________________ and __________________ rocks, there are two other types of rocks in the park: ____________________, which is hardened lava, and ___________, which is cemented volcanic ash.

6. Much of the rim of the volcano's _________________ in Yellowstone has been breached, or covered with volcanic products.

7. Remember, this is an active volcano, with a ___________________________ just several miles beneath the surface that is a reminder of the volcanic activity that was observed in the past and activity that we expect in the future.

8. Because of our Creator's design, even in a ___________________ world the forest is coming back quickly.

Discussion Questions:

1. What is volcanic tuff?

2. What effect does fire have on Yellowstone in terms of plant life?

3. Were larger animals able to escape from the fire and return to the park later?

Bonus Activity:

On a map, locate five other locations that have super volcanoes.

Chapter 3: The Column
Geologic Column

Fill in the blanks with words from the following list: **(one word is used twice)**

layer	Flood	fossils	log mats	sedimentary
Cenozoic	catastrophic	Creation week	buoyancy	Paleozoic
fossilization	sediments	vegetation	Precambrian	microbes
erosion	biblical worldview			

In order for us to better understand the geology of Yellowstone, we need to take a look at the Geologic Column.

1. The traditional belief from secular scientists is that the geologic column represents long ages of time, billions of years, starting with the ______________ and ending in the ________________ being laid down slowly and gradually without a global catastrophe.

2. Using the Bible as our history guide book, the earth is only about 6,000 years in age according to the genealogies. Since many of the layers are ________________, the Flood was the ________________ historical event that was the mechanism for quickly laying down each layer.

3. The ____________ we find in each _____________ were mostly buried over the year of the Flood.

4. Many ____________ rocks are those created during the week of creation, mostly on day 3.

5. There were likely forces acting upon them during the _____________ with the raising and lowering of the mountains and valleys to cause some changes, but largely they were the rocks made during the ___________________.

6. Because of the years before the Flood, we do see some evidence of the process of _____________ and some sediment layers, but there are very few fossils because the Flood, the prime mechanism for ___________________, had not yet come.

7. Secular scientists say that these early layers do not have fossils because life was too "simple" at the beginning, and only __________ had evolved.

8. The ________________ layers start with fossils of sea-floor creatures, which would have been buried first with the onset of the Flood. In layers above are fish fossils because they were the next to die. How does this make sense when you read the time-line of events regarding the Great Flood?

9. Above the fish and sea creature fossils are found land plant fossils and coal. When the Flood moved inland, it wiped out most of the ____________________. As trees were uprooted, they surely formed into some of the large floating ____________, similar to what we have seen at Mount St. Helen's Spirit Lake after the eruption.

10. Large reptiles could no longer survive. Their bodies may have floated on top of the water for a while, but eventually their weight and mass carried them to the ocean floor, where they were quickly buried by other ____________________. The natural sorting power of flowing water may have also contributed to distributing plants and animals in sedimentary layers through ______________ and other factors.

11. Most layers have a better explanation when viewed in light of a ______________________________.

Since there is not a full sequence of the column in any one location, this is consistent with the Flood account because water and currents will carry sand, silt, and mud to different places around the earth.

It's Flood action, not billions of years of activity.

Use the numbers between 1 and 7 to order the events into the sequence that created coal:

_______ the tree bark came off

_______ causing large deposits of peat

_______ which eventually turned into coal

_______ the bark sank to the bottom of the sea floor

_______ loosened logs rubbed together

_______ floating on the Flood waters

_______ when buried by other sediments.

Discussion Questions:

1. Read Psalm 104:8–9 and Genesis 2:10. How does this help our understanding of Yellowstone in terms of geology and geologic processes?

2. Why do secular scientists have to combine locations to get their full view of the geologic column?

3. How many years passed between the creation week and the Flood of Noah?

4. How does the Bible's account of creation and the great Flood explain the historical evidence of dinosaurs and their fossils?

5. How could the ability of land animals to move to higher ground during the first stages of the Flood impact their placement in the fossil record?

6. From which time period on the geologic column are the rock layers containing more coal and reptiles?

7. Fossilization of plants and animals happens quickly. It does not take long periods of time. What are some scenarios that could explain the lack of numerous mammal fossils and support the idea of them surviving longer during the initial stages of the Flood?

8. Why is the biblical explanation so much stronger for several reasons than secular explanations regarding the geologic record? (Clues: Where do "simple" life forms and fossilization relate to this?)

9. What are several reasons why we don't see fossils being formed today?

10. Which layer on the geologic column contains the last layers of fossils and represents the end stages of the Flood?

Bonus Activity:

Read Genesis 1:1–11; compare a secular geologic column to the creation week timeline.

Chapter 4: The Geysers

On the northwest side of the southern loop is a prize collection of geysers. It's a grcat place to get out and tour the hot springs.

Please note if the following statements are true (T) or false (F).

1. _____ Yellowstone hosts one-thirds of all the known geysers in the world.
2. _____ The close proximity of these geysers is because of the large magma chamber a few miles below the surface.
3. _____ The highest geyser in the world resides here: Steamboat Geyser. It can erupt up to 300 feet high, and eruptions are rarely unpredictable.
4. _____ A geyser's source is melted snow and rain, which eventually moves through the porous sedimentary rock.
5. _____ The water mixes with saline brine and heat from the shallow magma.
6. _____ The water rises well above the boiling point, but remains in a liquid state due to the pressure and weight of overlying water and rocks.
7. _____ The water can exceed 900 degrees.
8. _____ The silica in the water creates seals in the fissures and a plumbing system develops. When the pressure is sufficient, it forces the water up to the surface and a geyser results.
9. _____ There are a variety of minerals and species of algae in the hot pools. Microbes grow at different acidities. A pool's color cannot be used to determine its acidity. Orange is more acidic, and green is more alkaline.
10. _____ The organisms and enzymes can survive sulfur emissions similar to microbes living near thermal vents on the bottom of the ocean.
11. _____ Secular scientists suggest that if you look into the pools you imagine as it began billions of years ago.
12. _____ In reality, these heat-loving microbes are not simple or primitive, but extremely complex in their structure and biochemistry.
13. _____ Microbes, like all of creation, require an intelligent designer, the creator God of the universe. Evolved over billions of years? Not a chance!

Chapter 5: Yellowstone River
Yellow River and Hayden Valley

Fill in the blanks with words from the following list: (one word is used twice)

Yellowstone Lake	glacial	Hayden Valley buffalo
shoreline	fumaroles	Ice Age

1. Some of the hottest geothermal features here in Yellowstone are ________________.

2. The Yellowstone River runs through the park, with its source being ________________.

3. There were once millions of ______________ that roamed the plains and mountains of North America. They were killed off mostly by hunting, but now a strong herd exists in Yellowstone year-round.

4. ________________ is thought to have been covered by 200 feet of water when Yellowstone Lake extended through this valley after the retreat of the glaciers at the end of the __________.

5. Very little grows here because it was a lake bed where __________ clay lines the valley floor. Trees across the river mark the former ___________ of the lake.

Discussion Questions:

1. In which part of Yellowstone Park are bison normally found?

2. What effect has Yellowstone River had on rhyolite and soft tuff?

3. What causes the stained pastel color by the hot springs and fumaroles?

Chapter 5: Yellowstone River
Yellowstone Falls

Fill in the blanks with words from the following list:

ice dam — Missoula flood — tens of thousands of years

Grand Canyon of Yellowstone — fast-moving

1. There is an area in the park called the ______________________. It was cut by the Yellowstone River through basalt and ash and is one of the most well-recognized geologic features of the park.
2. Secular geologists thought that the carving of this canyon took ______________, but evidence is beginning to emerge that an upstream __________ may have failed, carving out this canyon in a very short period of time, maybe in even as little as a day.
3. The rock in the canyon is soft, so it would have been quite easily cut by __________ water.
4. There is similar evidence like this in eastern Washington with the ____________, at Mount St. Helens, as well as the Grand Canyon in Arizona.

Discussion Questions:

1. How does the steepness of the canyon walls help us realize there has been a recent cutting of the canyon?

2. How can we tell that Yellowstone was not formed by glaciation?

3. Explain the processes that form a U-shaped canyon.

4. Explain how understanding that this canyon has not been affected by glaciation helps us conclude that it was formed quite recently, after the Ice Age, only a few thousand years ago.

Bonus Activity:

What geologic features do Yellowstone, the Grand Canyon, and Mount St. Helens all have in common?

Chapter 5: Yellowstone River
Mud Volcano

Please note if the following statements are true (T) or false (F).

1. _____ Throughout the park there are some really cool mud pots, and they are some of the least acidic features in the park.
2. _____ In 1870, Mud Volcano was a miniature volcano. It spewed mud up to 450 feet in the air. Early explorers were awed by it.
3. _____ Hydrogen sulfide gas from deep in the earth is used by some micro-organisms as an energy source.
4. _____ They help convert the gas into hydrochloric acid, which breaks down rock into wet clay and mud.
5. _____ Escaping gases can't explode through the mud.

Chapter 6: Yellowstone Lake

At over 7,000 feet, Yellowstone Lake is the biggest and highest lake in North America. Some people are concerned because a bulge has been seen growing on the bottom of the lake. Some say it's going to erupt soon.

It's true that scientists have been closely monitoring Yellowstone Lake for volcanic activity and a growing bulge on the bottom of the lake. In studies of past eruptions around the world, a bulge in a volcano can be bad news. Look up recent volcanic eruptions from around the world; you may be surprised at how many of them occur!

Fill in the blanks below with words from the following list:

catastrophe	magma	dormant	mega-volcanism
Mount St. Helens	equalize	water	valley

1. Right before ________________ erupted in 1980, the entire north side of the mountain grew out in a large bulge, then the entire north side of the mountain slid into the __________ below and a nine hour eruption ensued.

2. Let's face it, the Yellowstone Caldera is not just __________; it's active, but very placid at the moment.

3. If the area were to become active in a big way, based on what we can infer from eruptions in the past, the effects could be devastating to the entire ________________.

4. This type is a remnant of the ________________ recorded in Genesis chapters 6–8, where the fountains of the great deep opened up, which included __________ and _______. We don't see it like that today because the earth after the Flood began to ________________ and the large-scale volcanic activity slowed down. But during that time, it was catastrophic.

5. Could the Yellowstone super volcano erupt again? Let's read 2 Peter 3:10 (NKJV) to find out: "But the day of the Lord will come as a thief in the night, in which the heavens will pass away with a great noise, and the elements will melt with fervent heat; both the earth and the works that are in it will be burned up." What do you think?

Chapter 7: Old Faithful

1. What was the average time between the eruptions of the famous geyser known as Old Faithful?

2. What are the two years in which earthquakes occurred that altered how regular Old Faithful has become?

Chapter 8: The Mammoth Hot Springs

On the northern loop in the northwestern corner, there is an amazing site where hundreds of feet of white terraces have been built up. During the time of the global Flood, about 4,350 years ago, large deposits of lime mud accumulated and were deposited in this area as limestone layers.

How do they continue to form?

Use the numbers between 1 and 9 to arrange in order the events of this terrace-building sequence:

____ Water flows across the surface.

____ Elements are left to form terraces.

____ Winter snows and summer rains bring a lot of water to the area.

____ The water and lime seeps out of the ground through fractures in the calcium carbonate beds.

____ Water is heated.

____ The water dissolves.

____ Water evaporates.

____ Water moves through the ground.

____ Lime is picked up by the water.

These terraces are always in a transitional state. Old ones stop forming while new ones begin. This is because the chemicals in the water frequently plug up the outgoing water, forcing it in a different direction where there is less resistance.

Some signs in the park suggest that Mammoth Hot Springs took 65,000 years to form, but at the current rate of deposition, averaging about eight inches a year, only a few thousand years would be required to make the terraces we see today. Now that is awesome!

Chapter 9: Petrified Forest

One of the most fascinating parts about Yellowstone is the Petrified Forest, as much of it is in thc Specimen Ridge area. There we find hundreds of upright petrified logs in 27 to 50 "layers of forests."

Please note if the following statements are true (T) or false (F).

1. ____ Secular scientists say that there were many forests here, one laid on top of the other, which took around 30,000 years in an ongoing cycle to create, but when you look at the evidence, it doesn't really make sense.
2. ____ A forest would continue growing when it would be covered by volcanic ash.
3. ____ Minerals would soak into the tree, and petrify it.
4. ____ The ash weathered into rocks and compost, then a new forest grew, was destroyed by the ash, and the process continued.
5. ____ Eventually the layers were exposed by glaciation, revealing what we see today.
6. ____ Yet even by the end of the Flood, some of the trees were still floating on the top of the Flood waters, even as the waters retreated from the new emerging land masses.
7. ____ As the logs became water logged, the heavy end, the root ball, sank in the water to the submerged land surface below. With the water currents still carrying the logs, they were completely lost.
8. ____ There was also a lot of continuing volcanic activity beneath the waters and on the emerging land surface, creating large explosions of ash that added to the accumulating silt.
9. ____ Because the logs were not sinking at the same time, they would come to rest on different layers of silt and ash.
10. ____ Eventually, the upright logs would be fully buried and the chemicals in the water and ash would petrify the trees over a period of a very long time.
11. ____ When the Flood waters fully receded, up to 50 levels of logs were deposited.
12. ____ If the logs were trees that grew at these locations, one would expect to find a well-developed root system under the logs, but they are non-existent in all of the mature trees at all 50 levels.

Bonus Activities:

Using the Bible as our history guidebook, with an earth age of only 6,000 years, the Petrified Forest of Yellowstone does not match up with the secular science of 30,000 years, but there is a different explanation. What is this biblical view? Find and note verses that support this explanation.

Chapter 9: Petrified Forest
What Is Petrification?

Please note if the following statements are true (T) or false (F).

1. _____ Petrification is actually a pretty complex process. It doesn't take that much time.
2. _____ Secular scientists used to say it took many years to petrify wood, but we can do it quickly in labs today.
3. _____ What you need to petrify is the presence of water saturated with elements such as lime or silica.
4. _____ We know from the caves around Yellowstone that these minerals are in abundance.
5. _____ When a log is buried and saturated with water containing these minerals, there is a chemical exchange that takes place.
6. _____ Researchers a few years ago did their own experiment by putting a piece of wood into a silica-rich hot spring at Yellowstone. One year later the wood contained substantial petrification.
7. _____ Even if we give an extended range, it could take as much as five hundred years to fully petrify a log. That's still in the biblical time scale.
8. _____ In fact, commercial hardwood is now being produced for flooring, which is man-made petrified wood.
9. _____ Many of the processes in nature secular science thought took ages we can now do very quickly. Coal and oil have been produced in less than a year.
10. _____ The global Flood would provide the right conditions to quickly bury and exchange the chemicals and heat necessary to produce petrified wood on such a grand scale as in Yellowstone.

Meteor Crater and Petrified Forest Worksheets

for Use with

Explore Meteor Crater and Petrified Forest with Noah Justice
(Awesome Science Series)

Chapter 2: Petrified Forest Introduction
The Grand Staircase and Noah's Flood

Fill in the blanks with words from the following list:

receded	geologic	remnant	fountains	volcanoes	mountain ranges
Genesis 6	Kaibab	Bryce	water	underwater	Chinle Formation
continental	ash	Grand Staircase			

1. The Painted Desert in northern Arizona is a ________________ of a vast dry lakebed, one of the many lakes that stretched across several western states after the Flood.

2. This area is part of the __________________________, a 10,000-foot-deep section of sedimentary strata, starting at the top of ________________ Canyon in Utah, and ending at the bottom of the Grand Canyon. Much of this area was formed through water, by the laying down of layers by water, and then eroded by water.

3. Much of this area is in the Chinle Formation, which is in the lower middle of the Grand Staircase. It's an amazing area, full of __________________ sites.

4. It is thought to have been formed over millions of years, but there's another story that better explains the features we see. Using the Bible as our historical authority, we see in ____________________ that God sent a flood to destroy the whole earth.

5. The ________________ of the great deep burst forth and it rained for 40 days and nights non-stop, and culminated in a year-long Flood. The bursting forth included water and volcanic activity as continents were broken apart. __________________ covered the whole earth for about a year.

6. As the continents moved under the water and slowly came to a stop, ___________________________ were quickly pushed up. Water from the Flood rushed off the continents, eroding valleys. But some water got trapped in large inland lakes between the mountain ranges and plateaus

7. Even after the Flood waters _______________, volcanic activity continued, then eventually slowed down due in large part to _____________________ movements minimizing and the earth equalizing after this great catastrophe.

8. The landscape of the entire southwest shows evidence that there were two large lakes that existed eastward of the ________________ Plateau.

9. Many creationists believe these waters breached and became the source of the catastrophic flow that drained quickly through the Kaibab Plateau to carve the Grand Canyon in a matter of just days. ________________ continued to erupt around this area, even after the Flood.

10. The ______________________________ was laid down, as evidenced by the volcanic flows of basalt on top of these layers.

11. Much of the Chinle Formation contains volcanic ______________ laid down by water. This volcanic ash came from volcanoes erupting, for the most part, ______________ during the Flood. When it mixed with sand and mud, this huge layer was deposited, along with logs and dinosaurs.

Discussion Questions

1. What is the Painted Desert?

2. How do volcanoes and floods impact the landscape?

Bonus Activity:

Find a map and see if you can draw a circle around the Grand Canyon, Meteor Crater, and the Petrified Forest. See if you can figure out how many miles apart they all are, and which might be closest to Mount St. Helens.

Chapter 2: Petrified Forest Introduction
About the Petrified Forest

Please note if the following statements are true (T) or false (F).

1. _____ In 1962 this area became a national park because of its unique features with 50,000 acres being set aside for the public to come and examine earth's history.
2. _____ Park signs say that this area was formed 225 million years ago, but is this accurate? This date comes from a belief in evolution and millions of years.
3. _____ According to the biblical record and genealogies, we can determine this area is only about 141,350 years old, formed at the time of the Flood.
4. _____ Everyone agrees these logs were transported in and by water, laid down here, then fossilized. But the mechanism for how they got here and how long the fossilization took is where the difference lies.
5. _____ You either think in long ages, a concept held by those who believe in creation . . . or just a few thousand years, according to the Bible, which evolutionists use as a time-line.
6. _____ First of all, the root balls on these logs are very small, or just absent, which scientists see as evidence that the original trees were ripped out of their original location in a cataclysmic event . . . think Genesis Flood.
7. _____ During the Flood, massive amounts of water rushed across the land, uprooting much of the vegetation, including large trees.
8. _____ The logs floated on top of the Floodwaters for a while, but eventually sank to the bottom and were buried quickly by the volcanic sediments.
9. _____ The bark has been stripped off, so something happened to cause the bark to be knocked off.
10. _____ During catastrophes, like Mount St. Helens, we saw something similar. After the 1980 eruption knocked down trees, logs floated below Mount St. Helens on Spirit Lake. They rubbed against each other, and eventually the bark fell off before the logs began being deposited on the bottom of the lake. During the Flood catastrophe we would expect nothing less.
11. _____ Where the bark fell was most likely different from where the logs rested, because the rock layers here, in the Petrified Forest National Park, do not contain any coal. Bark is commonly found in coal layers.
12. _____ We can see growth rings in the logs. They are large. The original trees would have grown up in a very healthy environment, which is what we would expect before the global Flood. The environment back then was definitely better than today, but still suffered the effects of the Curse.
13. _____ Due to volcanic action, the water was surely heavy with silica, and when the logs were buried, the silica-rich water would have petrified them.
14. _____ Why is this significant? Well, when the logs were buried, the carbon would have traded places with the silica in a chemical process called mummification, and the logs were quickly fossilized.
15. _____ Signs in the park will tell you it took a long time for these logs to fossilize. But with the right conditions, such as a global Flood, it could take less than a year. One lab has been able to duplicate this process in just days. Now that's repeatable science!

16. _____ Also, in Yellowstone National Park, scientists experimented with putting logs in silica-rich water and in less than a year substantial fossilization occurred.

17. Since this area continued to be underwater, even after the Flood receded, these logs stayed buried for a while. In a few years or less, the large inland lakes wore away at the limestone in the Kiabab Plateau, and a massive erosional event occurred, where the Grand Canyon was carved in just days. _________

Aftermath

When the Grand Canyon formed, some of these layers at Petrified Forest National Park were exposed. The exiting water eroded through these layers, exposing the petrified tress and creating the "teepee" geologic features.

Much of the desert floor is clay, which mostly came from volcanic ash during the Flood, but some probably came from volcanic activity after the Flood. But in some places like Jasper Forest, the stumps are buried in sandstone with pebbles. We call this a conglomerate layer.

When a rock layer has pebbles, it's usually a sign that fast-moving water was involved to round the rocks. This gives us another indication that water action was responsible for creating what we see here today.

It's interesting that we find entire logs here, and not just bits and pieces of them. Researchers realize that entire forests were swept away in one large event . . . the global Flood makes perfect sense!

Remember, don't sample or take any pieces of petrified wood. The park rules say no, and we need to keep things like this for future generations.

Discussion Questions

1. How does the evidence of trees being fossilized in less than a year support the view that the Petrified Forest was a result of the Great Flood?

Chapter 2: Petrified Forest Introduction
Other Petrified Examples

Fill in the blanks with words from the following list:

tidal wave   peat   catastrophic   landslide   Spirit Lake   silica   erosion

1. One way for us to verify our theories on ____________________ forest destruction, floating logs mats, and petrification, is looking no farther than Mount St. Helens in southern Washington State.

2. In 1980 Mount St. Helens had a huge eruption, causing the north side of the mountain to slide into the valley below, resulting in the largest ______________ in recorded history.

3. The landslide pushed into Spirit Lake and caused an 800' high ___________________ on the opposite hill. The wave was so huge that it pulled the entire forest of large, mature trees back into the lake in less than a few minutes.

4. The logs covered the top of _____________________ in one large floating tree mat, and many are still there today. The logs rubbed together and the bark sunk.

5. Over time, thousands of logs became water logged and sank to the bottom of the lake. Eventually, up to 20,000 logs sank to the bottom of the lake, and the layer of bark formed a layer of ____________________. Sonar tests and SCUBA expeditions have verified these facts.

6. If Mount St. Helens were to continue to erupt and fill Spirit Lake, all of the trees would be buried in peat, silt, and ash, which would naturally have _______________ in it. Eventually the trees would fossilize.

7. If exposed at some point through _____________, this area at Mount St. Helens would look much like what we see at Petrified Forest National Park, and even the numerous levels of apparent petrified forests at Specimen Ridge in Yellowstone National Park.

Even if secular scientists don't like the conclusion, this is evidence for catastrophic processes, like the Genesis Flood, that helped carve many of the geologic features we see around the world today.

What About the Teepees?

Fill in the blanks with words from the following list:

uniformity sediments deposited water erosion underwater monsoon layers

Noah's Flood

1. Driving along within the Petrified Forest National Park you'll see steep, cone-shaped hills, which are called "the Teepees." They are made up of volcanic ash, at the top of the Chinle Formation. Their shape is the result of ______________________.

2. In August it is __________________ season in Arizona, where afternoon storms will drop as much as 1–7 inches of rain in 15–30 minutes. The ash and clay erode quickly when this much rain falls over a short period of time.

3. The brilliant __________________ in the Chinle Formation indicate they were created by catastrophic means. How? Let's look at the evidence.

4. Indicators are that these volcanic ash layers were originally __________________ under water. It was not a local event, but huge, in order for the same colored layers to stretch out for extreme distances.

5. ______________________, just 4,350 years ago is an ideal candidate for a catastrophe on such a massive scale.

6. If it were a smaller event, then layers would be more sporadic. The different bands of color are due to different episodes of __________________ volcanic eruptions.

7. Every volcanic eruption puts out different materials, some coarse, some fine, so one layer is never the same as the next. When eruptions happen underwater, such variations are even more diverse because the water will also carry other __________________, which will mix in with the ash.

8. Slow and gradual processes would have laid down these deposits with no __________________. But if they were laid down by water rapidly, one after the other, the layers would be uniform over long distances and large areas.

Chapter 3: The Teepees
Rapid Layer Deposition and Erosion

Fill in the blanks with words from the following list:

earthen dam	hundreds	features	melted	periods	strata	processes
crater	airborne	erosional	ash layers	formations	Grand Canyon	
independent	alternating	debris				

1. Rapid layered deposits and erosion are not a fairy tale. When catastrophic ________________ are at work, amazing things can happen!

2. Back at Mount St. Helens, the 1980 eruption blasted the volcanic ash and debris to the north, piling ash layers on top of the initial landslide ____________________.

3. When finished, the valley floor had risen _______________ of feet.

4. Then in 1982, Mount St. Helens erupted again, but this time the __________________ contained a large amount of snow and ice.

5. When the eruption happened, the snow and ice ________________.

6. The water breached the large ____________________ formed by the crater rim and carved huge canyons near the crater, then continued to carve into the blast zone deposits below.

7. All of this __________________________ action happened in just a few hours.

8. One of these canyons has been called the "Little Grand Canyon" because of the similar erosional features we find in the ________________________.

9. These erosional ______________________ have allowed scientists to view the layers laid down by Mount St. Helens during past eruptions, as well as the 1982 eruption.

10. What they observed was different layers of ash, pumice and mud laid down rapidly, some by __________________________ processes, some by water processes.

11. What surprised scientists the most was that the ________________ showed independent multiple layering during the same eruption.

12. Even in the hurricane force spread of deposition, independent multiple layers were made ______________________________ between fine grain and coarse materials.

13. Mount St. Helens helped researchers realize how a catastrophe, like the Flood, can develop ______________________________. multiple layers in the strata, as well as large erosional features. This should make anyone question whether catastrophes have occurred in the past.

14. So when someone thinks that geological features took long ________________________ of time, this should be immediately questioned.

15. The ________________________ in the Teepees at Petrified Forest National Park could have been created quickly, over vast areas of the Southwest.

16. It's just here that the layers were exposed, then weathered away into these fantastic ________________.

The Bible can be trusted in what it says about catastrophic processes not long ago in earth's history, and researchers are realizing this the more they study catastrophes and geologic features.

Discussion Questions

1. Why was an event like the eruption at Mount St. Helens so important for researchers who support a biblical time-line of history?

2. What is erosion, and how can it impact the landscape during periods of floods or heavy rains?

Chapter 4: Dinosaurs
Fossilized Dinosaur Collection

Many dinosaur fossils have been found in Petrified Forest National Park, along with fossilized logs. This area obviously had a lot of catastrophe deposits during the Flood.

Signs in the park describe this area's past as a temperate rain forest where the dinosaurs lived. But there is little evidence to support this theory. Secular scientists point to the fossil fauna to support their story, but all evidence points to the Flood.

For instance, all of the strata are sedimentary rocks, laid down repeatedly by water. It was not an ancient forest floor. The deposits indicate quick deposition, including the ash beds in the teepee formations in the park.

The dinosaurs were not living here millions and millions of years ago, but were buried here quickly in sediments during the global Flood.

Discussion Questions

1. Dinosaurs are associated with a "millions of years" time-line of earth's history. How does the biblical account of the Great Flood give an answer to why dinosaur fossils are found around the world?

2. What are some clues that the Great Flood killed dinosaurs in the fossil record?

Bonus Activity

Choose a favorite dinosaur and see if you can find photos of a fossil online of this dinosaur. What do we scientists theorize about your favorite dinosaur in terms of where it may have lived, what it might have eaten, and whether or not it was a predator?

Chapter 6: Meteor Crater Introduction
About Meteor Crater

Northern Arizona is home to the Grand Canyon, Lake Powell, the Painted Desert, and many other geologic wonders. In addition to catastrophic processes due to the global Flood, there is one formation in the desert floor that has fascinated scientists for decades. It is a crater in the Kaibab Formation, the top layer seen at the Grand Canyon. The crater is 4,200 feet wide and 750 feet deep.

Fill in the blanks with words from the following list:

asteroid dust storm starved model Shoemaker acid rain meteorite

heat dust layers Yucatan Peninsula

1. Due to all of the volcanism in the area, it was once thought to be a volcano. But after scientific research, a whole new ______________ emerged.

2. Here at Meteor Crater, Arizona, a 150-foot-diameter ____________ slammed into the earth, creating this gigantic crater we see here.

3. Not long after the formation of the Grand Canyon, an ___________ weighing approximately 60,000 tons impacted the earth at around 40,000 miles per hour.

4. Because of what we know today about the earth's atmosphere and the ____________ experienced during entry, the asteroid was most likely much larger out in space.

5. In the 1960s, Dr. Eugene ________________ studied this crater, which led him on a worldwide search to find other impact craters. To his surprise, he found hundreds around the entire earth.

6. Some say that a meteorite like this a hundred times bigger crashed into the ____________________, causing the extinction of the dinosaurs. Think again.

7. They say the meteor created a giant ____________________, wiping out all of the plants, and thus wiping out the dinosaur's food source.

8. The dinosaurs around the earth would have ____________ to death by lack of food, then become buried in dust. But there's one major problem with this idea.

9. We don't find dinosaurs buried in ______________________. We find them in sedimentary rock layers, clay, sand, and dirt; this means flood action, not a gigantic dust storm.

10. Another proposed idea is a meteor crashed into the ocean, causing tidal waves and ____________, destroying life on the planet.

Discussion Questions

1. What is a meteor?

2. What is it about where dinosaur fossils are found that seems to indicate a meteor impacting earth didn't kill the dinosaurs?

Bonus Activity

Learn more about impacts. Put a towel on the table and then an unbreakable bowl half-filled with water on the towel. Drop a rock in the water. Re-create the impact, dropping the rock at different levels to see how that changes the impact on the water. Now, take the rock and use it to test impacts in a box of sand. What happens when the rock hits the water? Or hits the sand?

Chapter 7: Early History
Meteor Crater Early History

In the early 1900s, Daniel Barringer, a mining engineer and businessman, came across this large crater. Geologists told him it was volcanic because of all the evidence for volcanism in the area. But he had a different opinion.

Please note if the following statements are true (T) or false (F).

1. _____ He believed it was an impact crater from a meteorite, and began to perform research to show that it was. He didn't do his research for science, but for profit.
2. _____ If it was a meteorite, being a mining engineer, he knew that fragments, or even the meteor itself, would be worthless, making no money if sold.
3. _____ Meteorites are known to have high iron and nickel content.
4. _____ When iron ore is mined from the earth, it typically yields 65–70 percent iron. But when nickel ore is mined, you're lucky to get 1 percent.
5. _____ A meteorite will typically contain about 92 percent iron and 7 percent nickel. A large meteorite could be sold for a very large amount of money, so he began to try and find this one.
6. _____ He first thought he'd find it in the side of the crater.
7. _____ He dug down 700 feet in the middle but came up empty. Instead of giving up, he changed his plan of attack. Then he theorized the meteor came in at an angle. So he drilled into the rim.
8. _____ Barringer and his team went down 13 feet and found a few fragments, but not the big payoff of a meteorite.
9. _____ That was enough to convince him that it was a crater caused by a volcano, not a meteorite impact.

More Details on the Crater

What makes Meteor Crater so cool is that there are few large craters so visible to the public. There are a hundred or more of these craters around the earth, but we can't see them because they are eroded and obscured in a tropical rain forest or on the ocean floor. Meteor Crater is on the high desert of Arizona with nothing to hide it.

Please note if the following statements are true (T) or false (F).

1. _____ Meteor Crater is almost a mile wide and 750 feet deep. That's as long as two and a half football fields.
2. _____ With the meteorite coming in at approximately 40,000 mph, it caused a huge explosion on impact. Because of this great speed, most of the meteorite vaporized, leaving the crater and just small fragments scattered around.
3. _____ Signs at the crater say that this event happened around 500,000 years ago. But according to the genealogies in the Bible, the earth is only 60,000 years old, and God doesn't get things wrong.

4. ____ We know this event had to have happened before the water receded after the global Flood because it sits in a dry desert above the sedimentary rock. So this puts it more than 4,350 years ago.
5. ____ This area is part of the Grand Staircase, a 10,000-foot section of earth made of sedimentary layers laid down during the global Flood.
6. ____ Researchers have pointed out that after the Flood receded, two very large inland lakes, three times the size of Lake Michigan, were here in the early stages after the Flood, trapped between mountains and plateaus.
7. ____ In just a few years, the lakes wore away at the concrete in the Kiabab Plateau. When the water finally wore through, it carved the Grand Canyon in just a few decades, emptying the lakes and leaving this area as a high desert.
8. ____ Knowing that the Bible says the Flood ended around 4,350 years ago, and that these lakes existed after the Flood, then since the crater is sitting above the sedimentary layers, it might put this event at Meteor Crater between 3,000 and 4,000 years ago! But we can't be dogmatic about it. It's just an educated guess.

Chapter 8: Shoemaker
Shoemaker's Theory

In the early 1960s Eugene Shoemaker began to visit this area to study the crater. Studying impact craters was his passion. He studied craters at atomic test sites in Nevada and spent time at Lowell Observatory in Flagstaff to study crater impacts on the moon.

Fill in the blanks with words from the following list:

Chesapeake hypothesized craters evidence sea floor ashes meteorites

moon anomaly

1. He ______________ that meteor impact craters are much more numerous than anyone previously thought.
2. To him, Meteor Crater was not an ______________.
3. Based on his research, he proposed that there could be other ______________ around the world, and he began searching for them.
4. He discovered how catastrophic processes, like ______________ and nuclear bombs, create shatter cones, which is evidence for the shock of a meteor impact.
5. Then he used this ______________ to test other craters around the world.
6. He not only found impact craters on the continents, but also on the ______________.
7. Some of these craters have been found in ______________ Bay as well as on the Yucatan Peninsula.
8. He helped to train astronauts to study craters on the ______________. He was even a candidate to go to the moon, but could not because of his health.
9. He is the only person who has his ______________ buried on the moon.

Chapter 9: Dinosaur Death? Dinosaur Extinction Because of Meteor Impact?

Did a meteor impact cause the death of dinosaurs? Let's look at the facts.

Fill in the blanks with words from the following list:

numbers	carcasses	global flood	quickly	land-dwelling	dust storm
quick burial	Bible	population	water	tidal waves	perished

1. There are many ideas on how the dinosaurs died. Yet most of them seem to skirt around the ____________________ concept, even though most of the evidence supports it.

2. We know the dinosaurs were buried in sedimentary layers, which means they were buried in ____________________. Any other theory that does not involve water just doesn't stand up.

3. If the dinosaurs died on land, then their bodies would have decomposed or been eaten by scavengers. If they died in the sea, the same processes would have consumed their ________________.

4. But all evidence shows they were buried ____________ in sediments such as in sand and mud.

5. When we use the ____________________ as our history guide book, we see the mechanism of a global Flood.

6. This would have been the cause and the means for ________________ of the dinosaurs.

7. But the Bible also tells us that God brought two of every ____________________, air-breathing animal on Noah's ark, which included the various dinosaur kinds, estimated to be 50.

8. Any animal outside the ark would have ____________________ in the Flood.

9. As we already mentioned, some secular scientists have proposed an immense ____________________ caused by a giant meteorite hitting the earth, blocking out the sun, and dwindling their food supply. The dinosaurs simply starved to death in this scenario.

10. These impacts should have affected the whole world's animal ____________________, but they didn't.

11. Furthermore, dinosaurs were dying out after the Flood, which tragically reduced their ____________________, of course.

12. Other ideas say a meteorite hit the ocean near the Yucatan Peninsula, causing giant ____________ and acid rain to fall, killing off the dinosaurs.

Yet none of these ideas can adequately explain what we see, except for a global flood.

Yosemite and Zion Worksheets

for Use with

Explore Yosemite and Zion National Parks with Noah Justice
(Awesome Science Series)

Chapter 2: Yosemite Introduction
About Yosemite National Park

Fill in the blanks with words from the following list:

million deep valley alpine nature El Capitan glory order

California glaciers inspiration Rhode Island granite formation Biblical

Yosemite Valley

1. Yosemite is breathtaking from the minute you enter her cathedral __________ peaks rising several thousand feet above the valley floor.

2. It's no wonder author and champion for national parks John Muir said, "No temple made with human hands can compare with Yosemite, where God's attributes and ______________ is revealed" (Romans 1:20).

3. Muir believed that the best place to discover the true attributes of God was in ___________ and that God is always active in the creation of life, and thereby has continued to keep the natural ________ of the world (Colossians 1:16-17).

4. About the size of ________________________, Yosemite National Park covers 1,190 square miles and is full of valleys, giant cliffs, waterfalls, __________ lakes, and even active glaciers, with its highest point being over 13,000 feet.

5. Yosemite was first set aside by Abraham Lincoln on June 30, 1864, and given to the State of _________________to oversee. It was the ________________ for the first national park, Yellowstone. When the National Parks Service was organized in 1916, Yosemite was adopted in as a national park.

6. It is visited by over 3.7 million people, mostly in the seven square mile area of _______________________. From the valley floor, giant granite cliffs go straight up and have names such as Half-Dome, _______________, and Sentinel Dome.

7. As one looks around, questions come to mind about the formation of this amazing granite rock and how it was cut in such a _________________.

8. The granite is thought by secular scientists to have begun forming over 210 _____________ years ago.

9. After the ___________________ of the Sierras, the valley was mostly cut by the Merced River and finished off by large______________.

10. Obviously, these time scales don't match with the _____________ time scale of just a few thousand years.

Evidence has come to light to help explain the quick formation of the granites during the Flood of Genesis, as well as a model for the rapid formation of the massive glaciers and ice sheets of the Ice Age not long after the Flood.

What evidence? Let's take a closer look!

Discussion Questions:

1. What process do secular scientists think occurred to make the deep valleys?

2. How does the biblical story of Noah and the Great Flood offer a better explanation for these geologic features? You will be learning even more about the evidence in this study guide!

Bonus Activity:

Find Yosemite National Park on a map; see what photos you can find or a map of the park itself so you can find different geologic and natural features — rivers, mountains, valleys, etc. The National Park Service has websites for each of the national parks, but note, they often promote a non-biblical worldview for the age of the earth. With that caution in mind, they can be a great source to know more about the ecology and animals that make the park their home.

Chapter 3: Granite Formation
Evidence for Quick Formation of Granite

Please note if the following statements are true (T) or false (F).

1. ___The Sierra Nevada mountain chain stretches across the lower western United States. Its beautiful granite rocks and volcanic peaks make it an amazing place to visit.
2. ___Granite rocks make up a major portion of the continental crust. They are observed in many places on the globe.
3. ___These large areas of granite are called "Batholiths," which can range in size from just a few hundred feet, to thousands of miles.
4. ___Secular scientists have developed the idea that this large mass of intrusive igneous rocks formed far below the earth's surface in hot magma pools, and then they cooled, rose up, and tilted over many millions of years.
5. ___Because the story of evolution is so widely believed, it seems like there is a lot of authority in what secular scientists say.
6. ___You are probably familiar with the idea of biological evolution — how fish supposedly changed into frogs and then into me and you.
7. ___Well, secular scientists also talk about geological evolution — how the earth has changed over the supposed billions of years. They use the idea of uniformitarianism to explain this.

Discussion Questions:

1. How do the biblical genealogies in Genesis help us calculate the age of the earth?

2. There are some within the Church who try to mix long ages, millions and billions of years, with the Bible. The Bible is pretty clear on the age of the creation and the Flood . . . only about 6,000 years based on the days of creation and the genealogies found in Genesis. Secular scientists reject this age because they dismiss the Bible out of hand, but why do you think some Christians follow secular dating of the earth instead of a biblical one? (Hint: Science is built on theories and hypothesis; what if you assume they are right because science says so, versus it being observable or proven through the evidence available for testing today?)

Bonus Activity:

See how many verses you can list in the Bible that speak of God's creation of the world, animals, and people. (Hint — not all of them are in Genesis!)

Chapter 4: Cutting of Yosemite
The Receding Floodwaters: Cutting of Yosemite

Fill in the blanks with words from the following list:

Cliffs	Muir	mountains	erosion	hypothesize	global	chains
Merced	continents	catastrophe	Glacier Point	water		

1. As you enter through Yosemite Valley, the granite __________ rising above 3,000 feet impress the viewer with the grandeur of God.
2. From ______________________, a visitor can view Yosemite Valley below. Without question, something big cut this valley.
3. Secular scientists claim that as the _____________ were pushed up over millions of years, the ____________ River flowed a little bit faster and cut the valley.
4. It's hard to imagine this little river cutting this huge valley. It just doesn't make sense based on what we know about the principles of __________________. There's just not enough __________ flow.
5. In addition, what was the source behind the Merced River to allow it to erode this much land? It just doesn't exist, except in the case of a massive _______________, like the global Flood.
6. In the 1800s, John _______________ explored the park and was the first to ________________ glaciers were involved in eroding Yosemite.
7. It's true that glaciers cut and smoothed a valley that already exists, but a valley as big as Yosemite requires something much larger . . . like a ______________ flood.
8. Near the end of the Flood, the _______________ were beginning to rise. Mountain ____________ were being pushed up, and valleys were being formed.

This catastrophic activity that was moving continents subsequently caused the waters to accumulate in the ocean basins and huge sheet erosion planed off new top layers of the earth (Genesis 8:3–5).

As there was less water flowing toward the ocean, the erosion continued as the water flowed in channels, cutting giant canyons, water gaps, and valleys.

Evidence for this massive water erosion is everywhere across the earth, including Yosemite Valley. But the erosion of this famous valley wasn't over yet. The Great Ice Age was to come next and it also left its marks.

Chapter 5: One Ice Age
Evidence for Quick Formation of One Ice Age

Researchers point out indicators in Yosemite where glaciers once filled this area almost up to the top of Half Dome. The large fields of ice flowed slowly downhill on top of the granite, leaving behind evidence of their movement by marks showing their direction of travel. This is reminiscent of other areas also carved out by glaciers like the granites of Sweden.

From our past study, we learned the granites were formed during the time of the Flood, so the ice came after, within the last 4,350 years.

Secular models say the last ice age peak ended 18,000 years ago, but that doesn't match the biblical time scale.

Please note if the following statements are true (T) or false (F).

1. ___Secular scientists have over 60 ideas on how the Ice Age formed, but each idea is riddled with problems because they don't allow for a catastrophic event like the Flood.
2. ___Most simply appeal to a cold globe, but that doesn't make an ice age, just a cold earth. To get an ice age, you need two ingredients: hot oceans and cold continents.
3. ___Yet, the Flood provides a good model for quick formation of the Ice Age, and the even quicker melting of those same ice sheets.
4. ___The biblical Flood was not worldwide.
5. ___Under the waters, the surface of the earth remained completely unchanged by cataclysmic event (Genesis 7:21–22).
6. ___The Bible tells us that the Flood started with the fountains of the great deep bursting forth and massive amounts of rain fell for 40 days and nights and the flooding continued until the 150th day (Genesis 7:11–12, 7:24).
7. ___After day 150, the water began flowing into the sinking ocean basins as mountains rose higher. Most of the water used in the deluge continually ran off the continents and into the ocean basins (Genesis 8:3; Psalm 104:7–8).
8. ___Because of the massive friction due to continental movement, volcanic activity, and other factors, the waters at the end of the Flood were very cold. And this is one of the keys to an ice age: warm oceans to cause lots of moisture to evaporate into the air.
9. ___This wet air is then carried over the continents by storms.
10. ___The clouds and the ash in the air cool the climate and don't allow the snow to melt, even in the summer. Over many years, the ice builds into massive sheets and then you have . . . an Ice Age.

Through ocean core samples, researchers suggest that the oceans right after the Flood warmed to an average of 90 degrees, with some places near 120 degrees.

The average today is about 62 degrees, so that is a dramatic difference! We know that warm water — really warm water — is a hotbed to build large hurricanes and storms.

In some computer models, these storms could drop up to 30" of rain an hour. If correct, as they moved inland and up across the mountains, this could also contribute to some of the massive amounts of snow.

Chapter 5: One Ice Age

Fill in the blanks with words from the following list:

snowfall	Flood	equalize	systems	summers	ice	model
oceans	meadows	Ice Age	precipitation	winter	large-scale	accumulation
Silver Lake	sunlight					

1. The world record for snowfall in recent history is held by ________________, Colorado, where 76 inches fell over 24 hours back in 1921.
2. The conditions right after the Flood would have dropped a lot more snow than this with few breaks between ________________.
3. But ________________ alone wouldn't cause an Ice Age. You would also need cooler summers for the snow and ice to persist.
4. There is evidence all around the world for ________________ volcanic activity at the end of the Flood and several hundred years after as the catastrophic geologic processes were slowing down.
5. This volcanic activity would have released a continual flow of ash into the atmosphere, causing very cool ________________ as the ash reflected ________________ away from the earth.
6. As the earth began to ________________ from the effects of the Flood, the building of glaciers would have slowed down as the ____________ cooled and the ash became less in the atmosphere.
7. Eventually the ____________ building would have stopped and began melting.
8. Some creation scientists have calculated that with the conditions right after the Flood, the full extent of the ________________ could have reached its zenith as soon as five hundred years after the____________.
9. The amount of ____________________ would have built up huge ice deposits in the Sierra Nevada mountains and filled in Yosemite Valley with ice in just a few hundred years.
10. Because secular scientists refuse to include a catastrophic global Flood model into their research, they are left with trying to figure out an Ice Age _____________ using current conditions . . . and the numbers just don't work.
11. From the valley to the high ______________ at Yosemite, the current yearly mean temperature ranges between 25 to 90 degrees. The precipitation falls mostly in the ______________ with nearly dry summers.
12. Average ______________ is only 65.6 inches a year, and all of that is gone by the summer. For an Ice Age to build, you need snow staying on the ground year round.

For 3,000 to 4,000 feet of ice to develop, the conditions would have to have been drastically different in the past. Secular scientists still don't have a valid theory to explain these conditions, but the Bible does.

Science, it's awesome!

Chapter 6: Melting Ice Age
Evidence for Quick Melting of the Ice Age

Use the numbers between 1 and 4 to order the events into the sequence that help reveal one ice age instead of many:

___As the fresh glacier water emptied and melted into the oceans, the colder fresh water would have sat on top of the warmer salt water for a while closer to the coasts, slowing down the evaporation and heavy storm development.

___As more of these ice dams broke across the continents and brought more fresh cold water into the oceans over several decades, the seas eventually equalized into the current temperatures.

___But after the fresh water eventually mixed in to the oceans, the surface temperature would have warmed back up and the large storms would have continued, only lesser in strength.

___These intervals likely produced features that might look like multiple ice ages from the advancing and retreating of the ice sheets, but there was only one Ice Age with some minor forward and backward movement of ice.

Chapter 8: Zion Introduction
About Zion National Park

Fill in the blanks with words from the following list:

Patriarchs	Arizona	crossbeds	Angels	sandstone	habitation	national park
flowing	bottom	references	cathedrals	Utah	understanding	challenges
Zion	Grand Staircase					

1. The western United States has many amazing and remarkable geologic sites, but one that is a favorite among visitors is Zion National Park in southern ______________.

2. Zion is a part of the ________________________, a 10,000-foot series of sedimentary layers stretching from the bottom of Grand Canyon in ______________ to Bryce Canyon in Utah.

3. Zion is a relatively small ________________ compared to many, at 229 square miles and only 15 miles long.

4. The __________ layer of Zion is the same layer found at the top of Grand Canyon. So Zion Canyon could have been eroded before Grand Canyon. Zion became a national park in 1919, but ______________ of the canyon by Native Americans goes back thousands of years.

5. The first permanent European-American settler in the canyon called his new home "Zion" because he felt, "A man can worship God among these great _____________ as well as in man-made church" (Psalm 2:6).

6. Many of the formations in the park have biblical ____________________ as the Great White Throne, the Three _______________, and _____________ Landing. They were named by Frederick Fisher, a Methodist minister.

7. In addition to the canyon, the east side of the park contains amazing rolling hills of sandstone __________________ in an area called "Checkerboard Mesa."

8. Standing over 2,000 feet tall, these are some of the highest sandstone cliffs in the world. The crisscrossing patterns are due to the layers being deposited in ____________ water.

9. This valley has clear signs of the _______________ being formed by water, then eroded by water.

10. ______________ includes nine separate geologic formations, which secular scientists think cover 150 million years of earth's history.

11. Yet those who believe in long ages have some major ___________ explaining the features we see at Zion, because a proper _______________ of the evidence doesn't mesh with their story.

Instead, astute researchers recognize the catastrophic formation of the layers and cutting of this enormous canyon. It wasn't millions of years, but perhaps just weeks. Zion is a towering example of geologic action due to catastrophic flood action, the biblical Flood of Noah's day (Genesis 7).

Chapter 9: Formation of the Layers

Please note if the following statements are true (T) or false (F).

1. ___As one meanders through the giant canyon, the walls of sandstone are enormous. The largest layer is the Navaho Sandstone, which is up to 2,200 feet thick.
2. ___This layer is known to cover over 140 square miles in the southwest.
3. ___The source of its sand grains can be traced to the northern Appalachians over a thousand miles away.
4. ___Secular scientists have suggested several ideas on how this sand moved across the continent. One is that an ancient river transported it from the northeast to the southwest, but no evidence exists for this grand waterway.
5. ___Another idea is that they were blown across by wind. But how did they miss being deposited in the Midwest as well? When the details surface, this idea is dashed and broken "against a big wall."
6. ___Without a global flood model, secular scientists have a major challenge explaining how this much sand was moved this far away. But with worldwide catastrophic conditions during the Flood, the sandstones are easily explained.
7. ___When sand is blown, the backside of a dune is typically more than 30 degrees. Also, if water moves sand, the backside of the dune is less than 30 degrees. This can be observed around the world and can be tested in the lab.
8. ___None of the crossbeds in the Navajo Sandstone have angles less than 30 degree, therefore we can deduce that they were moved by water.
9. ___Only one event in earth's history had the water and speed to transport 140,000 square miles of sand — the global Flood as told in the Bible.

Chapter 10: Cutting of Zion Canyon

When you tour through Zion Canyon, one question keeps entering your mind . . . how was this canyon cut? Whatever the process was, it had to work its way through thousands of feet of sediment.

receding	obstruction	remnant	small	water	receding	younger
gradually	gravity	cut	forces	collapsed	higher	features
hanging	directions	V-shaped	canyon	Virgin River	arches	intersection
Big Bend						

Fill in the blanks with words from the following list:

1. Secular scientists attribute Zion's erosion to the ____________________, which flows through the bottom of the canyon, over millions of years.

2. Just a casual glance reveals that the river flows too ________________ to have caused this much erosion. The river doesn't even reach the canyon walls.

3. What is more, as one looks deeper into the canyon's ______________, it becomes evident this place was __________ by a lot of water over a very short period of time . . . just days or hours.

4. One of the strongest evidences researchers use for a young cut _____________ is how steep the canyon walls are. If the canyon was formed over a long period then ______________, erosion, and other natural _______________ would have caused these steep walls to collapse into the canyon in various places.

5. As a result, this valley would have been ___________ with lots of debris. But there is very little debris on the canyon floor, which is a sign of a much ______________ canyon.

6. In several places around the park, there are huge ____________ high in the canyon walls. These in the sandstone will eventually erode away and fall to the canyon floor.

7. If the Virgin River took millions of years to carve this canyon, then these should have ______________ and eroded away a long time ago, but they are still there. In addition, the park is full of ______________ valleys.

8. These valleys are formed when massive amounts of water are moving through an area at high speed, cutting the landscape quickly from many ______________, generally due to flooding conditions. Some valleys are ____________ than others and spill into the lower ones.

9. When the floodwaters go away, the ______________ of the upper canyons into the lower canyons are left hanging.

10. Zion is full of these giant hanging valleys. A huge amount of ____________ moving together at the same time would be required to create these features.

11. If ____________ streams, over long periods of time, were responsible for carving the hanging valleys, then they should have be eroded down or be completely gone. Yet they remain as a ________________ of a global Flood when the _________ waters raced across the entire continent.

12. One of the features in the park secular scientists point to as evidence for slow and gradual erosion by the Virgin River is a place called "________________." They say if catastrophic processes were at work, the ____________ Flood waters should have just eroded this feature away.

13. We know that when flowing water finds an ______________, it usually finds a way around it, and it is usually found to be anything harder than the layer being eroded. The water finds a weaker point and goes that way. This is exactly what happened in Big Bend. And this is consistent with a biblical view.

Discussion Questions:

1. Why can't the Virgin River be the cause for Zion Canyon's formation?

2. How does the steepness of the canyon walls help to prove a younger time line for the canyon being formed?

3. If the canyon were older, would there be more or less debris on the canyon floor?

Bonus Activity:

Find photos of different valleys — see if you can find four examples of V-shaped and U-shaped ones.

Mount St. Helens Worksheets

for Use with

Explore Mount St. Helens with Noah Justice
(Awesome Science Series)

Chapter 2: Mount St. Helens Introduction
The Grand Staircase and Noah's Flood

Fill in the blanks with words from the following list:

Toutle	Cascade	active	Flood	Mount Rainer	tectonic	sixty
dormant	steam	Spirit	virgin	magma	earthquakes	high
British Columbia		Washington				

1. The Pacific Northwest in the United States is an amazing collection of pristine coastland, lush valley farmland, _____ desert, and the __________ Range.

2. The Cascade Range stretches from Northern California to southern _________ ______________. The range was pushed up during the later stages of the _____, and contains about a dozen volcanic peaks, averaging about 10,000 feet.

3. Most of the peaks are thought to have formed not long after the Flood, when the earth was still equalizing from the massive ____________shifting.

4. Eventually much of this volcanic activity slowed down in about 500 years, but a few volcanoes remained __________ while others went _______.

5. The recent increase of population centers around these peaks has drawn concern from scientists. The 14,400-foot dormant ______________ poses a huge threat of catastrophic destruction if it should erupt.

6. Another dormant volcanic peak was Mount St. Helens in southern ___________ State. This area was a pristine scenic wonderland, with tall and beautiful ___________ forests, as well as deep blue mountain lakes.

7. Youth camps and mountain cabins lined the shores of _________Lake and the _________ River north of the mountain. For decades, brave mountaineers would climb the 9,677-foot summit for a specular view.

8. But in March 1980, the mountain started to awaken. At first small ______________ began to rattle the countryside.

9. Over the next __________ days, there were over 12,000 earthquakes, each increasing in size. Scientists knew the sleeping giant was about to wake up.

10. Then in early April the first ____________ explosion penetrated the summit and a hole appeared in the snow. As earthquakes increased, scientists believed that __________ was slowly working its way up to the surface.

In early May, a bulge began to appear on the north side of the mountain. It was estimated to be growing at five feet a day. Like a giant balloon, the pressure was growing, and the danger level of a large-scale eruption appeared to be imminent.

On the morning of May 18, 1980, at 8:32 a.m., an earthquake registering 5.1 on the Richter scale signaled the eruption.

Chapter 2: Mount St. Helens
About Mount St. Helens

Please note if the following statements are true (T) or false (F).

1. ____This earthquake caused a giant landslide as one-half of a cubic mile of summit slid north into the valley below, creating 25 square miles of new landscape.
2. ____The avalanche did not contain rock, snow, or glaciers.
3. ____When the landslide slid off Mount St. Helens, three-fourths of it went into the Toutle River Valley, raising the valley floor 600 feet.
4. ____As the landslide came into the valley, huge chunks of the mountain, called "hummocks," stayed intact.
5. ____Half of the landslide traveled northeast and spilled into Spirit Lake, causing a 260-foot tidal wave across the water, washing up onto the opposite hillside and totally destroying the old-growth forest there.
6. ____The new landslide material also permanently raised the level of the lake more than 2,000 feet above the pre-eruption level.
7. ____The landslide contained 13 glaciers from the top of the mountain, which were buried in the landslide and eventually covered with ash.
8. ____After this, 1,300 feet of the mountain disappeared in the landslide, a massive steam explosion came, spreading across the landscape to the north.
9. ____This steam explosion went lateral, uplifting 150 square miles, causing the new-growth forest to look like toothpicks laying on top of each other.
10. ____Usually when a mountain first erupts, the explosion goes straight up, and Mount St. Helens did something similar since the first explosion went straight out to the south.
11. ____For the next nine hours the mountain released the equivalent of 40 million tons of TNT blast energy. That's equivalent to 33,000 Hiroshima-sized atomic bombs, or one atomic bomb a second.
12. ____The dirt and magma cloud spread across the eastern part of Washington State. Cities like New York and New Orleans were turned from day to night in a matter of days.
13. ____As the buried glaciers in the landslide debris heated up, they eventually exploded, causing large pits in the landslide material.
14. ____All of the melted snow and ice caused a mudflow down the Toutle River.
15. ____The mudflow carried ash, pumice, and rocks, tens of miles down the valley, all the way to the Mississippi River.
16. ____Shipping lanes in the region were never shut down because of the debris clogging up the river.
17. ____Down the Toutle River valley, bridges were totally gone. Logging camps were destroyed. Houses were washed away. The devastation was massive.

Discussion Questions

18. Having seen this type of blast for the first time, scientists were now able to find 300 other volcanoes on earth that they were able to explain based on the observations from Mount St. Helens. What other volcanoes are known to have erupted the same way Mount St. Helens did?

19. What descriptions in Psalm 104:23 that are hinting at volcanoes speaks to the power of God?

20. Is Mount St. Helens considered to be a massive volcanic event or merely a small to average? Was the volcanic activity in places like Yellowstone larger or smaller than that of Mount St. Helens? (Hint: One of these eruptions at Yellowstone was estimated at 2,500 times larger in its destructive force.)

Bonus Activity

See if you can find news reports about the eruption online and read the way the event is described; here are some details to help you:

- The day after the eruption, there were 57 people dead and the devastation was beyond description.
- In recent memory, no one had ever seen such catastrophic destruction.
- President Jimmy Carter even flew over the area a few days after May 18 and described the blast zone as looking like the surface of the moon.

But in the destruction came a research opportunity for volcanologists and creation scientists. As the eruption began to be studied, it revealed how many geologic features around the world could have happened as a result of major catastrophes, in particular, the global Flood.

Chapter 3: Ape Cave

Fill in the blanks with words from the following list:

Ape	hot	Pacific	lava	caves	Kilauea	"Bigfoot"
Oregon	formations	peaks	extinct	south	flow	Mazamma
foresters						

1. The Cascade Range has a wide range of volcanic _________. Some have remained dormant, some have gone _______, while others remain active.
2. Crater Lake in southern Oregon is the remnant of Mount __________, which imploded a few thousand years ago, and its ash can be found all around the ________Northwest.
3. Another imploded mountain is at Newberry Crater in central ________, and like Yellowstone, is considered active because of its ________springs.
4. On the _________ side of Mount St. Helens great ancient lava flows can be seen at the timberline level. Those who explore this area will see some great geologic __________.
5. Most volcanoes have ______________. At Mount St. Helens, it is ________Cave. It's one of the longest in the world at almost 12,000 feet.
6. The origination of the name for Ape Cave is somewhat unclear, but some think it goes back to alleged ________ sightings in the area back in 1924, where a "big ape" was seen.
7. Others say it's due to the ________________and loggers from many years ago who were referred to as "Brush Apes." Another group says it was named after a Boy Scout troop called, the "apes" back in the 1950s.
8. This lava cave was formed when the lava _________ cooled on the top, but the hot lava still ran underneath. Eventually it got lower and lower and left the cave altogether.
9. This type of cave formation can be seen at active volcanoes like ___________in Hawaii.

Use the numbers between 1 and 6 to order the events into the sequence that created lava caves:

____a cave is left

____the top hardens and a lava tube forms

____eventually the lava stops flowing

____hot lava flows through a trough

____the top of the trough cools because it's closest to the surface

____the rest of the lava keeps flowing

Discussion Questions

1. The upper Ape cave is about 1.5 miles long and climbs up 400 feet where cavers hike over 27 boulder piles and scale an 8-foot-high lava fall. How old do scientists think the tunnel may be? Has the cave been created by only one lava flow or several over time?

2. Why do researchers think it is unusual for Mount St. Helens to have produced a lava tube like this one?

3. How can the lava tube help us understand how Mount St. Helens formed after the Great Flood? What does all of this say about the power and awesomeness of God?

It's pretty amazing to think you can hike the same path where molten hot lava flowed down Mount St. Helens. This is a place to truly explore and see God's handiwork!

Science, it's awesome!

Chapter 4: Rapid Strata Formation

The eruption on May 18, 1980, at Mount St. Helens was impressive. When the north side of the mountain slid into the valley below, it created a gold mine of research material for creation scientists like Dr. Steve Austin who studied many of these formations.

Please note if the following statements are true (T) or false (F).

1. ___The valley below was covered with 6,000 feet of landslide deposits.
2. ___The nine-hour eruption laid huge deposits of ash and pumice on top of the landslide.
3. ___On June 12, not more than a month later, another major eruption put an additional 25 feet of deposits on earth's newest landscape.
4. ___For almost two years the mountain went quiet. Then in March 2015 there was another major eruption, which melted ice and snow that had collected in the crater.
5. ___Large amounts of water mixed with ash and pumice.
6. ___This created a "lahar," or flash flood, which came down off the mountain and laid another layer on top of the past layers. The strata — layers of deposits — were forming quickly.
7. ___As the mudflow reached the north side of the blast zone, the water came to an obstacle and began to pool.
8. ___The water eventually eroded through the obstruction and carved some amazing canyons on the valley floor.
9. ___As scientists looked at the layers in the canyon walls, they saw what could have been interpreted as individual volcanic events based on the way the layers appeared. Scientists call this "building a sequence."
10. ___Secular scientists have held that geologic layers take long ages to form because they hold to the idea of *uniformitarianism,* which means lots of change over long periods of time.
11. ___At Mount St. Helens, these layers were formed in a matter of hours and there was direct observation of their formation.
12. ___For example, the 25-foot June 12 layer was formed in just three hours.
13. ___The layer was formed when the mountain erupted and a hot pyroclastic flow rushed down the northern flanks of the volcano at an amazing speed. Because of the erosion, we can see this layering in the canyon walls today.

Discussion Questions

1. If one were to explore these canyons formed during the events of Mount St. Helens, would you have any idea how quickly these layers had formed without knowing the date of the eruption?

2. Why is the following conclusion incorrect: A uniformitarian scientist might have guessed there were many eruptions over long ages because that's the main idea that has been taught over the last one hundred years, and the layers look similar to rock layers they assume took long ages to form.

3. What is catastrophism, and why has it been ignored or discounted by the secular scientific community despite evidence that supports the concept? (Hints: Noah's Flood and "long" ages)

4. Why can't secular scientists admit the global Flood of Genesis occurred?

Chapter 4: Rapid Strata Formation

Fill in the blanks with words from the following list:

Secular	homogenized	sedimentary	water	pyroclastic	seconds	canyons
elevation	particles	fine	lamina	geology	evidences	receding
landscape	micro-thin lamination					

1. Uniformitarianism has dominated the study of ____________ for the last 150 years as God's Word has been rejected.

2. As scientists looked closer at the June 12 layer, they discovered fine and coarse materials had been laid down as separate layers called ________. What was amazing was that these had been formed in the ____________ flow in winds moving over 100 mph!

3. It would be logical to assume that catastrophic processes would just mix up all of the material into one big ____________________ deposit with no distinct layers. But here at Mount St. Helens, just the opposite has been observed.

4. In these layers, there is a feature called ________________ where there are coarse and ________ layers just millimeters from each other.

5. This amount of detail is remarkable, because it has been proposed by __________ scientists that such features take long ages to form, not minutes or__________. Yet this is what has been observed to have happened at Mount St. Helens.

6. In other ____________ across the earth, we can observe many similar features, such as in the lower sandstone layers at Grand Canyon.

7. Secular scientists have proposed that all ___________ layers in the Grand Canyon were formed by seas coming and going over millions of years. As the seas persisted, sediments from the waters made the layers as _______________ settled out of the water. Eventually, the seas left and the land was pushed up to its current _______________.

8. There are many ______________ we can look at to show that these layers were not formed by seas over millions of years. Indications are that these layers were formed very quickly by _________ during the year of Noah's Flood.

9. As the Flood waters moved around the earth, large amounts of sediments settled out, forming layers of sediment. When the Flood waters were ____________, they continued to dump sediments on the _______________ in very short order.

With the observable evidence left after the eruption of Mount St. Helens, we now realize how quickly fine and coarse layers can form, given the right conditions. It doesn't take millions of years to form these types of layers . . . just the right catastrophic conditions, such as we would expect during the Flood.

Mount St. Helens provides a miniature laboratory for the study of particle stratification and concepts related to the Global Flood.

Chapter 5: Rapid Erosion

Discussion Questions

1. Secular scientists point to canyons around the world, like Grand Canyon and Zion Canyon, proposing that the small rivers in the bottom of the canyons carved what we see today over millions of years. What assumptions on the age of the earth are they making to explain this secular theory?

2. A proper understanding of the evidence after actual observations of the rapid canyon formation at Mount St. Helens led many researchers to conclude that to carve canyons of large magnitude you need a lot of water in a short period of time, not the small rivers over millions of years. Why was it so important that scientists be able to see these events actually take place in understanding the evidence left behind at other locations?

3. Contrary to what most scientists might think, it isn't the river that carved the canyon, it was the canyon that formed and provided a passageway for the river to flow through. Explain how this concept undermines the secular model regarding the role of rivers in canyon formation.

Chapter 5: Rapid Erosion

Please note if the following statements are true (T) or false (F).

1. ___ If one were to walk through this canyon using the uniformitarianism model of long ages, one would imagine it took tens of thousands to millions of years for the north fork of the Toutle River to carve this canyon.

2. ___Yet we know from eyewitness accounts that it happened very rapidly.

3. ___While we may not have direct eyewitness accounts of how the major canyons around the world were formed, we do however have the Bible, God's eyewitness testimony that gives us a framework by which we can look at these other canyons and features.

4. ___The biblical record and subsequent models based on what we have observed from other events we did witness are irrelevant to understanding these other canyons.

5. ___At Mount St. Helens we were not able to see the landscape before these canyons were formed and we know the events and mechanisms that laid down the strata and carved the canyons through them.

6. ___Because of the events at Mount St. Helens, even many secular geologists are junking the idea of millions of years for the formation of Grand Canyon and are thinking in terms of catastrophe.

What type of catastrophe would have cut the Grand Canyon and other huge canyons around the world? You would have needed a lot of water over a short period of time. There's only one event recorded in human history that is the key to accomplish this . . . the Flood of Noah's day as recorded in the Bible.

Fill in the blanks with words from the following list:

eroded receding ideas trapping dam gaps creationists Kaibab Plateau

1. There are a number of __________ of where the water came from to carve the Grand Canyon. Some creationists have developed the idea of two great lakes behind the ______________.

2. The _______ formed by the plateau was breached and _____________ the canyon in a matter of days as the lakes drained rapidly.

3. These lakes would have been left from the _____________ floodwaters as valleys and plateaus quickly rose at the end of the Flood, ____________the water in these huge lakes.

4. Other ____________ have suggested that the Grand Canyon and many other canyons around the world were formed when the floodwaters were receding across the land, cutting huge _________in the landscape.

Using Mount St. Helens as a laboratory for studying catastrophic processes helps us to realize the incredible impact the worldwide Flood had on forming the earth's features we see today!

Lava Cast Forests

Just to the south of Mount St. Helens is a fascinating feature called the Trail of Two Forests. Here is your opportunity to learn a little of the history of this feature!

Bonus Activity

Research the Trail of Two Forests online and see if you can find images of it. Create a poster with some of the images you find and the following information points:

- About 2,000 years ago, when the lava flow came through here, there was a tree standing in this exact place.
- As the lava flowed around it, it hardened enough against the cool wood to make a form right there.
- And then the wood vaporized through the heat.
- Whatever was left just rotted away, leaving a hole.
- When the lava came through, not all the trees were upright. Some of them fell down and created lateral tunnels all across the area.

Since that time, a new forest has grown on top of the lava flow, hence, the trail of two forests. An easy-to-use walkway has been built for us to see this great volcanic feature.

Chapter 7: Logs and Log Deposits

Please note if the following statements are true (T) or false (F).

1. ___When Mount St. Helens erupted on May 18, 1980, the largest landslide in recorded history slid down into the valley below.
2. ___Three-fourths of the landslide barreled into Spirit Lake, causing an 860-foot wave across the lake.
3. ___That huge wave washed up on to the surrounding hills and into the old-growth forests.
4. ___When the water came washing up from the lake, it pulled down about one billion trees with it.
5. ___The landslide also displaced the lake so that the present level is 2,000 feet above the level it was before the eruption.
6. ___Mountain cabins and lakeside camps were buried in a matter of weeks.
7. ___As the eruption stopped and scientists were able to get into the blast zone, an amazing sight began to be revealed. Some first thought Spirit Lake was gone because they couldn't see it, but it was covered with logs.
8. ___As the lake emerged, secular scientists were excited to know this would be their first chance to study how logs would behave after this catastrophe.
9. ___What they found would reinforce the way they had always thought about the development of petrified forests.
10. ___Since 1980, the number of logs on top of the lake has steadily increased dramatically.

Where Did the Logs Go?

Fill in the blanks with words from the following list:

root balls	living	resin	sink	waterlogged	mapped	mystery
soaked	absorption	bark	upright	sonar	interpret	species
Spirit Lake						

1. After the eruption, scientists studied how the logs became_______________.
2. Because the __________________ at the bottom of the logs were denser, they began to sink first, forcing the log to stand __________, halfway in the water.
3. As the logs __________up more and more water, they began to ________ to the bottom of the lake.
4. What also became evident was certain __________ remained floating, while others disappeared beneath the waters.
5. After careful study, it was discovered that some tree species, such as Noble and Silver Fir contained less____________.

6. Resin slows the ___________ of water, so those logs with more resin would stay afloat longer.
7. The other curious thing about the logs still floating is that their ___________ had been stripped off. Where did it go? The only logical explanation is that it went to the bottom of the lake.
8. These processes were new to scientists, so they began to study in this ___________ laboratory.
9. What excited them the most was how what they found could help them __________ other geologic sites around the world.
10. With many of the logs and all the bark gone from the top of the lake, they felt the real ____________ lay below the water.
11. After getting the right permits, they first took a small boat to ________________ with a sonar tow fish. They went back and forth among the giant logs and ________________ the bottom of the lake.
12. What they found was amazing. The ___________ map showed as many as 10,000 small and large logs standing straight up on the bottom at various levels in the sediments.

Then Dr. Austin did what any good scientists would do . . . further investigation.

Fill in the blanks with words from the following list:

park sign	subterranean	ash	chemical	long ages	Yellowstone	petrify
log mats	pressure	fissures	burrowing	rings	great deep	uniformitarian

1. The tree _____________ all matched in size, there was no evidence of _______________ animals, and the ash in the soil mostly came from the same eruptions.
2. The petrified forests at ________________ were formed by catastrophe in very short order. The_____________, which told of multiple forests over millions of years, was taken down.
3. The ____________________ explanation of the evidence just didn't hold up.
4. At the beginning of the global Flood, as described in the Bible, the rains came down and the fountains of the ______________ were opened.
5. It is believed these fountains were ________________ water and volcanic__________. During the Flood, water pushed across the land, ripping up forests across the landscape.
6. Some of these logs were buried instantly, but many floated to the top of the water, creating giant floating _____________, like those seen at Spirit Lake.
7. In various places around the world, these logs would have begun to sink to the bottom of the waters and buried quickly in sediments and ________.
8. With the immense ____________ from above, the heat from below, and the right _____________ mixture in the ash, the logs would have petrified quickly.
9. Some secular scientists have told us that it takes ______________ to petrify wood, but actually it really doesn't take that long. Experiments have been performed in the lab that found that logs can _______________ in less than a year.

Bonus Activity

- Want even more ways to learn about the event? *The Case of the Missing Mountain* by Kim Jones is a fun and activity-based learning tool for children studying Mount St. Helens. With clues to decode, mysteries to solve, and special project activities, it is a great way to extend your study of this incredible event!

Chapter 8: Peat and Coal Formation

The catastrophic events at Mount St. Helens have not only helped explain petrified forests, but also the formation of our large coal beds.

Please note if the following statements are true (T) or false (F).

1. ___The coal deposits around the world can be hundreds of feet thick and provide fuel for heating and electricity generation.
2. ___The coal deposit layers are usually very glassy and rough.
3. ___Secular scientists have developed the idea that these massive coal deposits were formed slowly over billions of years in fresh water oceans.
4. ___In the secular model of how coal forms, the logs fell from the forest over long periods of time, and were buried in the antiseptic waters of the swamp.
5. ___Over millions of years, a thick, spongy layer of broken plant material developed, called "peat."
6. ___This peat layer never was buried by other sediments such as clay, mud, and sand. It eventually turned into coal.

The hypothesis of coal forming from peat in ancient swamps sounds like a good story, but there are some significant challenges with this secular idea. This is especially true when you study the quality of swamp peat and peat beds.

A Different View

Using numbers between 1 and 5, see if you can put together this important sequence of events of how the giant coal beds we find today were formed in the Great Flood:

___Because of wave action and winds, this vegetation is thought to have clumped together on the surface, creating giant floating log mats.

___The sediments applied pressure to the peat to make coal in very short order.

___ After much of the peat was deposited, the subsequent flood waters would have laid sediments on top of the peat.

___ As described earlier, the Flood would have ripped up much of the vegetation as the waters prevailed on the land.

___ Just like at Spirit Lake, the logs would have rubbed together, and the bark would have fallen off and sunk to the bottom of the sea to form a layer of peat.

Discussion Question

The type of coal we find in the giant coal beds is very smooth and glassy, and by natural means, can really only be formed by rapidly laying down tree bark under water. How does this show that it was a catastrophic event rather than swamp action over millions of years that formed it? (Hint: be sure to get to the "root" of the matter!)

Does coal take long ages to form? No, it doesn't. Labs today are making coal in just a few weeks, so it does not take long ages like secular scientists would propose. All you need are the right catastrophic conditions.

Using Mount St. Helens as our miniature laboratory, the Bible as our history book and ultimate authority, and the Flood as the catastrophic process, the giant coal beds can easily be explained using this floating log mat model. And God's Word is the key to unlocking the mystery.

Chapter 9: Biological Recovery

Fill in the blanks with words from the following list:

nitrogen	spring	sections	hundreds	Legionella	side	bacteria
scat	pocket	blast zone	Lupine	toxic	plankton	nutrients
design	oxygen	organic	hibernation	reintroducing	algae	alders
biological recovery						

1. Much of Spirit Lake was considered a ________ brew of volcanic gasses seeping up from the lakebed.
2. Because of all the _______ material now in the lake, it became a hydrogen bubble, stinking of methane. Almost all of the _________ in the water was depleted.
3. The temperature of the water had risen 20 degrees C. _______________, which causes Legionnaires' disease, was also found in the lake.
4. Very little could live there except for ____________ and a little bit of ___________. It was quite a mess.
5. At first glance, every living thing had been destroyed. Many scientists thought it would take ___________________ of years for any life to return. But because of God's amazing ____________ in nature, life returned much sooner than expected.
6. The lumber companies decided to replant their _____________ of land with new trees, but the Forest Service decided to let their land grow back on its own. It became a living laboratory on ________________________ after a natural disaster.
7. Because it was _________________ when the mountain erupted, there was still snow on the ground. Many animals were still in ________________. Baby trees were also under the snow.
8. As spring turned into summer, the _____________ gophers came to life, breaking up the soil and spreading the seeds into the blast zone.
9. Elk would eat plants outside the ________________, come in and leave their scat on the ground.
10. The seeds in the _____________ would start growing as plants. In the same way, birds also carried seeds in.
11. But much of the soil was heavy in ________________ because of the ash. Most plants don't grow well in soils filled with this chemical element, but some plants do.
12. _______________ began growing like crazy across the landscape. It is able to eat up the nitrogen and develop the soil into a friendlier place for larger plants.
13. Eventually ______________ and conifers began to grow. Since 1980, a young forest has grown up in the blast zone.
14. In Spirit Lake the bacteria went to work by eating up the toxic chemicals. The _______________ put oxygen back in. And within five years the water was back to its original quality.

15. Fish were found in the lake again. Some could have stayed in ____________ streams, but a majority of the comeback was due to fisherman _______________ trout into the lake.

16. Within a few years, it was known that some fish were reaching lengths of 25 inches, much due to the good __________________ provided by the ash and sediments from the eruption.

Discussion Questions

1. Why is the fast recovery after the eruption so important to those of us who believe in the Bible as God's history book? As you consider the question, keep these two important points in mind:

   - God is good and a good designer of our planet. Remember, God has designed our earth to recover from catastrophe much quicker than secular scientists used to think possible.
   - He may use His finger and make the mountains smoke, but He also desires that there be a quick restoration of what was destroyed — we see the goodness of God even in this sin-cursed and broken world.

2. How does the recovery at Mount St. Helens also give us a glimpse into the quick recovery of our planet after the global Flood?

3. Compare and contrast the eruption with the global Flood. Think about each catastrophic event in terms of scale (how much area was affected), length of time during the event itself, why the events occurred, and clues to biological recovery occurring afterward.

4. Create a time-line of the events of the Great Flood from information we are given in the Bible. (Hint: Just over a year after boarding the ark, Noah and the animals walked on to dry land.)

Bonus Activity

Earlier we learned about lupine and how it actually helped other plant to recover by what it did to the soil. See what you can find out about lupine — a picture, and even how it works to remove nitrogen from the soil. Then see if you can find other amazing plants like this that can make a difference in the environment allowing other plants or animals to flourish!

John Day Fossil Beds Worksheets

for Use with

Explore John Day Fossil Beds with Noah Justice
(Awesome Science Series)

Chapter 2: Fossil Beds Overview
About John Day Fossil Beds

Fill in the blanks with words from the following list:

expedition	monument	plants	coastline	fossils	volcanic	John Day
lava flows	Oregon	fissures	basalt	three	thousands	

1. The state of ____________________ is divided into four major areas.
2. The landscape includes pristine ___________, rich valleys, cascade mountains, and the high desert located to the east. Much of the high desert is covered in ____________ that is thousands of feet thick.
3. The basalt comes from a series of lava flows originating from numerous ___________ in southeastern Washington and northeastern Oregon.
4. Some of these _____________ traveled as far as the Pacific Ocean, located a few hundred miles to the west.
5. Some of the basalt has been removed, creating large open areas ______________ of feet deep, revealing underlying layers of sandstones, shales, and _____________ deposits.
6. These sedimentary beds have colorful shades of tan, green, and red. They often contain ____________, mostly of ___________, but a few animals as well.
7. Some of the most specular sites where this sediment has been exposed were set aside for the national __________________. Spread over __________ different locations, over 100,000 visitors come every year to experience the outstanding geology and scenery.
8. This part of Oregon was named after John Day, a hunter and trapper who came west on an _________________ party in 1812. He and a partner became separated from their group and were raided by Indians.
9. This happened where the ________________ River enters the Columbia. The river meanders through eastern Oregon and the region.
10. The monument was established in 1975 and encompasses about 14,000 acres.
11. It is easy to make your way around the monument on highways and gravel roads. If you're passing through this area, it's really worth a visit.

The Two Models

As one tours around the monument, all of the interpretive signs explain the area's history from a secular, evolutionary perspective.

To many, the word "evolution" means our world has come about by natural processes through long ages of geological and biological evolution. This religious worldview says that every living thing came about by chance from a single cell, billions of years ago.

- This idea combines biological evolution and chemical evolution, trusting by faith that chemicals on the early earth formed living organisms that were able to reproduce and change into more and more complex organisms. It is the unifying religious theme for the secular worldview in biology and paleontology, and is connected to most of the other sciences as well.
- This worldview assumes that life spontaneously arose from non-living matter, and that mutations and natural selection drive evolution toward increasing fitness and complexity. However, none of this would be possible without a prior event — the big bang. So something popped into existence from nothing and eventually formed stars . . . and squirrels. These are all important aspects of the religion of secular humanism, which dominates society and science today. This religious worldview sets man as the ultimate authority and denies the existence of God.

The correct worldview to examine these amazing features through is special creation and global destruction by a Flood, as presented in the Bible.

- This view starts with the Bible which is 100 percent accurate and places God and His Word as the authority, not man's opinions and ideas. From a biblical worldview, we live on a planet close to 6,000 years old based on the genealogies in the Bible.

Discussion Questions

1. Which view do you personally feel is right? Why do you feel that is the best perspective to explain these landscapes?

2. God is a God of order and He has created the universe to operate on orderly principles. Look up the definition of scientific laws — can you see how God's orderly creation is revealed?

Bonus Activity

Many of the great scientists in the past understood that God upholds the universe in a particular way, making scientific study possible. These scientists had no conflict with their belief in God and the Bible and their study of God's created order. See how scientists like Keppler, Boyle, Mendel, Farady, and Newton have greatly advanced our scientific understanding.

Chapter 3: The Formations

Please note if the following statements are true (T) or false (F).

1. ___As you drive through the large canyons and valleys, it becomes evident that only one geologic formation can be seen in the steep walls.
2. ___From the top of the canyon to the bottom, there are at least ten formations.
3. ___The top formation contains sediments and glacial deposits, and the bottom of the canyon contains marble, schists, and limestone.
4. ___Creationists and evolutionist disagree that these rock layers are real and have distinctions; they agree on the timing of laying down these layers.
5. ___Secular geologists use this column to describe the long ages of the evolutionary history of the earth. Each layer has a set of index fossils that are used to identify it.
6. ___For the most part, biblical creationists also use the same column, but they interpret what they see mainly as layers being formed during the Flood in quick succession, not over long ages.
7. ___Arguably, there is no one location on earth that has the full column stacked one layer on top of the other. Scientists piece it together by comparing multiple locations around the world. Let's learn more about the process of layering.

Fill in the blanks with words from the following list:

thousands	magma	post-flood	pillows	quickness	lava beds	short
during	evolutionary	balance	local	slumping		

1. There is strong evidence that heavy volcanic activity was happening __________ and after the period when the floodwater was moving into the ocean basins as the earth came to a ____________ point, so this event could have happened then.
2. One reason why scientists think the basalt may have been formed after the floodwater moved off the continent is that when magma comes to the surface at the bottom of the ocean, it has a tendency to form ____________. Yet most of the basalt here has not done so.
3. So we can't be absolutely certain, but we still have a pretty narrow window — much less than the millions of years in the ______________ view.
4. The pillowing features seem to be related to the quantity and ___________ at which magma exits to the surface.
5. With the massive quantity of basalt here, we can be fairly certain that the earth was putting out very large amounts of ___________ at the end stages of the Flood.
6. The other challenge with this lava being a ____________ deposit is the fact that erosion of this area is so massive. The Flood is the only logical event which could have provided the means to carve the basalt ________________ of feet deep.
7. Yet there just isn't evidence that can be interpreted as a ___________ flood event in this part of Oregon.

8. As you enter the John Day River Valley near the small village of Antelope, it's easy to see the lines of contact between ________________. They are straight and level.
9. This evidence is consistent with a ____________ period of time between one flow and the next, and the whole sequence accumulated quickly. There were not hundreds of thousands of years between flows — otherwise, we'd see erosion.
10. There is also a large amount of sliding and ____________ in the layers. So the John Day beds were not thoroughly consolidated when formed, but were still soft or plastic when the bending action happened, then they hardened.

During the Flood, massive tectonic activity was happening below the waters to release the magma and ash.

Discussion Questions

1. What evidence in the conglomerate layers and in the remains of volcanic activity point to a catastrophic event?

2. There are a large variety of fossils at the John Day Fossil Beds. The Bible tells us that God sent the Flood to destroy all land-dwelling, air-breathing life, except for Noah, his family, and the animals on the ark. Do the fossils found here support that?

Chapter 4: Formation Erosion
Erosion of the Formations

Thousands of cubic miles of material have been removed from this area. Where did it go? There's no significant accumulation of sediment anywhere between here and the Pacific Ocean.

Fill in the blanks with words from the following list:

shelves	erosional	movements	gaps	excavation	Pacific Ocean	sheet erosion
properties	ocean basins	canyons	elevation	Mount St.Helens		

1. For hundreds of miles up and down this river valley, the ______________ of the Columbia River Basalt, and other layers has been enormous.

2. Water erosion has amazing ____________ and can cut through just about any substance, even basalt. In 1982 at ________________, a mudflow caused by immense amounts of water came out of the crater and carved ___________ hundreds of feet thick, including a lava flow thought to be 500 years old.

3. After day 150 of the Flood, the Bible tells us that the waters were indeed receding. Below the surface, the remaining ___________ of continents and changes in ______________ were happening.

4. Mountains were being pushed up, valleys were being formed, and what is now the __________________ was sinking lower.

5. These massive geologic changes started moving the water off the continents and into the __________________. During this time, giant ______________ was moving sediments off the top layers and dumping them on the continental ____________ in the ocean.

6. As the water decreased, it began to carve canyons and _____________ in the mountain chains and plateaus.

7. It is believed that this massive ________________ event was responsible for carving the canyons in the John Day Fossil Beds.

The absence of significant accumulations of erosional debris and sedimentary deposits between here and the ocean suggests that most of the material eroded from these great canyons was dumped into the Pacific Ocean. The Genesis Flood could provide sufficient water to accomplish this task quickly.

Bonus Activity

What is a "fossil bed"? Are they common or very rare? And why are they so important to helping us learn more about the dinosaurs, plants, animals, or insects buried there and the kinds of events that helped to form the fossil beds?

See if you can find five different kinds of pictures of fossilized leaves.

Have any dinosaurs been found at the John Day Fossil Beds?

Chapter 6: The Palisades

Fill in the blanks with words from the following list:

bedding	lahars	avocado	jungle	high desert	secular	leaves
eruptions	particles	ash	sorted	volcanic mudflow		conglomerates

1. The Palisades are composed of ________________ with angular rocks and boulders. They are thought by ________________ scientists to have formed 44 million years ago in a series of volcanic mudflows called __________.

2. Some scientists claim that this area was vastly different at one time . . . that this dry ________________ used to get 100 inches of rain every year and was surrounded by volcanoes in a near-tropical ____________ rain forest.

3. They say this area was near-tropical rain forest because most of the fossil ____________ in the rocks came from a rain forest environment. Similar modern ones are found in Central and South America, southeast Asia, and Africa. These include palm trees and ________________.

4. But strong ________________ could be an indication that some of the layers could have been ______________ by water, or dropped from the air into the water.

5. There's no question about these cliffs being formed with water. Secular scientists say it was a ______________________, but a stronger case is made that these cliffs were made during the Flood.

6. Ash from these ________________ would have generated the air fall ________________, which dropped into this formation. Leftover vegetation from the beginning of the Flood would have mixed in with the ____________ and created these layers.

Petrified Logs

Please note if the following statements are true (T) or false (F).

1. ___The Arch Trail works its way up to and along the tall cliffs. Fossils are abundant here. There are even petrified logs in the deposits. Some are horizontal, and some are even vertical. The horizontal logs all point in the same basic direction.
2. ___ The petrified logs cannot be explained by a mudflow, but they can be explained in light of a large-scale Flood.
3. ___ We know from places like Mount St. Helens how logs can end up vertical in a formation when the root ball becomes soaked with water and drops to the bottom, then is covered by sediment and fossilized.

4. ___ Based on lab experiments, petrified logs can be formed quickly — it doesn't take millions of years.

5. ___ What's interesting about these two conglomerate layers in the cliffs is that they are separated, probably by an air fall tephra deposit or an ash deposit.

6. ___ It could be that this was not just one deposit, but a series of deposits, enough time for a small ash deposit to be made before the next wave of rocks and mud came through this area.

7. ___ Another indicator that these cliffs are old is the lack of extensive erosion on the cliffs. Erosion can be caused by rain and wind, but also the freezing and thawing seen here every season.

8. ___ The current rate of erosion cannot be determined based on observations. There are large boulders on the ground, but nothing significant to suggest millions of years. The cliffs could have remained steep if long ages had passed.

9. ___ We also know this area regularly experiences small earthquakes, which would also shake loose rocks from the rock face. Over long ages, this cliff would have been shaken apart and eroded.

10. ___ Furthermore, using all of these principles of erosion, based on current rates and conditions, this area is about 1,600 feet above sea level and would have been eroded down to sea level in 12 to 15 million years. So even using the secular time scale, these cliffs could not be 44 million years old.

11. ___ In addition to plant fossils, animal fossils are also found here. These include tiny four-toed horses, huge rhino-like brontotheres, crocodilians, and meat-eating creodonts. In other places around the monument they have also found saber-tooth tiger fossils.

12. ___ These animals still live here today, and most of them lived in a tropical environment, just like the plant fossils found here. What we find here is just a remnant of what lived before the Flood.

13. ___ Another piece of evidence, or lack of evidence, that this area was not an ancient forest is the absence of a thick layer of leaves, humus, and debris on the forest floor before the mud flow came.

14. ___ If a jungle-like environment was here, there should be thin layers of fossilized forest. Yet the only evidence used to support the presence of ancient forest is the fossilized flora in the rocks.

15. ___ There is no evidence these fossil plants actually grew here. What we do know is water was involved in bringing them here. A much better model is they were carried in during the Flood, deposited, and then fossilized.

The catastrophic Flood described in the Bible helps us provide a much better explanation of what we see here at the Palisades.

Chapter 7: The Painted Hills

Fill in the blanks with words from the following list:

root balls	destruction	floodplain	flora	vegetation	permineralized coal
pre-Flood	rainfall	evidence	drying	paleontologists	

1. Secular____________________ suggest that thick forests grew here, mainly of hardwoods. They also say _______________ was approximately 40 inches a year.

2. To explain the formation of the Painted Hills using secular ideas, three different lakes and ____________ environments are postulated, with ____________ and time gaps in between, over some 40 million years.

3. This time is needed to allow for the development of the __________, one above the other and sufficient time for erosion and the _______________ of the previous environment. Yet there are some major challenges with this idea because the _______________ of long ages is not readily apparent.

4. If a forest grew here, we should see tree limbs and maybe small _____________, but none of that is found in these layers. We should even find some _________ from the decaying trees, but none of it is there.

5. There are black smears in the beds, but they are _____________ wood, not significant compared to what is expected from an ancient forest growing here.

6. We also know that it only rains an average of 14" a year in this region, but the fossil _____________ found here grows in a more lush environment. These types of fossils are what we would expect to find from _____________ environment.

Discussion Questions

1. What do the presence of flying creatures in the fossil record at John Day reveal about the scope of this catastrophic event?

2. What clues are found in the preservation of the leaf fossils that also support a quicker event, versus forest activity over millions of years?

Chapter 8: Gable Creek Conglomerate
The Cretaceous Gable Creek Conglomerate

Please note if the following statements are true (T) or false (F).

1. ___Along Highway 207 near Wheeler is a fascinating geologic formation in the John Day Fossil Beds, called the Cretaceous Gable Creek Conglomerate, dated around 30 million years old by secular scientists.
2. ___The problem for creation scientists is that there is about 30 million years missing between the Eocene Clarino Formation, and this one — by their own calculations!
3. ___Clearly, this wasn't formed over millions of years, but by the Flood over a short period of time.
4. ___The Gable Creek Conglomerate reveals starling evidence. It was created with catastrophic processes and contains the Hudspeth Marine Sandstone intermixing with the Gable Creek Conglomerates.
5. ___You can see the direction of flow found in the rocks. One came from the southeast, the other from the northeast.
6. ___The Hudspeth sandstone contains marine fossils while the Gable Creek Conglomerate contains a variety of rock types, including quartzite and granite. Layers cover 70 square miles and are 9,000 feet thick in some places. _____
7. ___Since quartzite mountains occur in Oregon, these rocks would need to have been rolled from somewhere else. The nearest source for those quartzite cobbles is the Rocky Mountains of Idaho and Montana, several hundred miles away.
8. ___Based on studies by creation scientists, only currents over 50 mph in slurries of over 100 feet deep could have move these rocks.
9. ___Secular geologists cannot easily explain these amazing geologic features. The Hudspua formation is sandstone and contains marine fossils. This means that these rocks were formed underwater in a large flood event.

Chapter 9: Picture Gorge

Please note if the following statements are true (T) or false (F).

1. ___The Picture Gorge Basalt covers over 10,000 square miles and makes up 2 percent of the total volume of the Columbia River Basalt.
2. ___These flows most likely came from fissures, or cracks, in the earth's crust, which can be observed down the road from Picture Gorge by Fossil.
3. ___Secular geologists will often tell you the small John Day River cut this canyon over long ages. Yet with some closer examination of the evidence, this may not be the case.
4. ___ Just the fact that the gorge has near-vertical walls is an indication these steep walls are not so old.
5. ___If the gorge existed for thousands of years, current rates of erosion would have rounded, and removed much of the material forming this canyon.

6. ___Water usually finds the lowest point to travel through. After a careful analysis of the geology and landscape, the John Day River should have flowed past this area and through Keyes Creek Pass just to the west, which is about 1,000 feet lower. But it went through this area we now call Picture Gorge.

7. ___In geomorphology, this feature is called this a "water gap." Many water gaps are seen around the world where water has cut through elevated areas.

8. ___Some secular scientists say that the land was pushed up after the gap was made. It's pretty easy to see this is not the case.

9. ___If the ground was pushed up, then you would see broken and cracked layers, which you just don't see. If the ground rose too quickly, it would dam up and go the other way. If the dam broke, you should see evidence of flooding, which you don't see.

Discussion Questions

1. Explain how it was not the small river that caused the canyon, but the canyon that now makes a path for the river in an event like the great Flood.

2. One of the geologic markers near Picture Gorge says, "The sharp, steep walls of Picture Gorge suggest a sudden cataclysm. . . ." But the sign continues, "and not the slow, relentless forces that actually shaped it." What portion of that sign supports a creationist perspective, and which supports a secular perspective?

Chapter 10: Sheep Rock

Fill in the blanks with words from the following list:

violent colors cap Mascall formation localized sediments

diatoms disjointed connected

1. The bottom layers are the John Day Formation, mostly made up of ash. The different _______ represent the different stages of the _____________. The lower is red, the middle is greenish, and the upper is a buff color.

2. Secular scientists say the John Day Formation was made 18 to 39 million years ago. But because of the ___________ nature of the mammal skeletons and the _____________ nature and varying thickness of these beds, slumping and reposition of these sediments has likely occurred underwater. Put simply, the Floodwaters were active and ___________ while this layer was being deposited.

3. What's cool about Sheep Rock is the ______________ at the top. After all the ash layering, the top is basalt from the Picture Gorge Basalts.

4. Sheep Rock used to be ____________________ to the other formations here in the valley, but the receding Floodwaters carved around it and left it standing alone.

5. After the massive lava flows covered the John Day Formation ____________ of ash were laid down during the late stages of the Flood in the Pliocene-Miocene _______________ Formation.

6. Secular scientists say this layer was laid down 12 to 15 million years ago, but what isn't admitted too often is this layer contains ____________, tiny one-celled plants found in oceans.

Discussion Questions

1. The top of the Mascall Formation is smooth. Why does that make it unlikely to have been impacted over millions of years?

2. What kinds of fossils are found in Blue Basin?

3. Can you discover why the best time to hike Blue Basin is in the morning or evening? (Hint: see what you can find about weather.)

Chapter 11: Mammals of John Day Fossil Beds

Please note if the following statements are true (T) or false (F).

1. ___Among fossils found here are several catlike carnivores, over a dozen different members of the dog family, horses, several different rhinoceroses, and a number of different camels.
2. ___The visitor center and published materials say these animal fossils evolved from common ancestors, yet most of these mammals fit into the family categories of ancient mammals.
3. ___The secular idea of evolution generally says the ancestors of modern animals were "simpler" and the modern more "complex."
4. ___If true, we should see a chain of animal fossils connecting these links.
5. ___Yet when we look back through the fossil layers, the earlier mammals are just as complex. Evidence of evolution from simpler animals is abundant.
6. ___Of course, we expect the complexity of early animals since each kind of animal was made on days 5 and 6, with intricate details by God.

Discussion Questions

1. Are animal kinds as noted in the Bible the same thing as species? Can you give an example of an animal kind that has many species?

2. Why don't creationists have a problem with what is known as natural selection?

3. How does a global flood explain the presence of fossils from environments different than that in which they were found?

Bonus Activity

Research a creationist named Ed Blyth. What idea did he propose that is credited with being discovered by Charles Darwin?

Chapter 12: Goose Rock
Goose Rock and Pediments

Discussion Questions

1. What are pediments — why is their presence near Goose Rock such a mystery?

2. Can pediments be explained by receding floodwaters? If so, how?

Yellowstone & Grand Teton Worksheets

for Use with

Your Guide to Yellowstone & Grand Teton
(True North Series)

Terms to Know (Write the definition in the space below.)

caldera

plateau

magma

evolution

Short Answer Questions

1. At what elevation will you find Yellowstone National Park?

2. What is the name of the mountain range just south of Yellowstone?

3. The Yellowstone Grand Teton Ecosystem is over ____________________ acres or about ______________ square miles.

4. Yellowstone has over ______________ geothermal features within the park.

Discussion Questions

1. Read Genesis 1:31, and Genesis 6:5-8. It must have been very difficult for God to have destroyed something He had created and designated as "very good." How is this example of God's righteous judgment helpful to us today?

2. Read Genesis 7:16, and John 10:9. In your discussion, compare these two verses. What has God provided for everyone who places his or her faith in Jesus Christ?

Activity

1. Research and plan a mock vacation to Yellowstone. How many miles is it from where you live? How long would it take to drive there? Would you camp out or stay in hotels? How much would food and fuel cost? Plan an itinerary and budget for the trip. Who knows, maybe one day you will get to actually take your "mock" vacation!

	Yellowstone	Sections 1–4 Pages 10–29	Day 93	Worksheet 2	Name

Terms to Know (Write the definition in the space below.)

uniformitarianism

creation

geologic time

sedimentary rocks

Noah's flood

worldview

Short Answer Questions

1. Traditionally, the earth's sedimentary layers and fossil beds were thought to represent what?

2. What have uniformitarian thinkers discarded in their theory of millions of years of geologic time?

3. Complete this sentence: God created life __________ ____ _______________.

4. Some critics of a global Flood suggest there is not enough water on the earth to cover all the mountains. Yet today, if the topography of the earth were flattened to form a perfectly round sphere, the planet would be covered by water over _____________ miles deep

5. Fill in the blank spaces in the chart with the following statements:

Discussion Questions

1. Read Genesis 1:28. Discuss ways that we can be good stewards of our national parks today.

2. Explain in your own words what you think Johann Kepler meant when he penned the words "thinking God's thoughts after Him."

3. According to Darwinian evolution, each innovation starts with a small, random, genetic mutation, but unless the resulting component has a survival value, it is not passed on to future generations. Discuss some of the problems with this theory.

Activity

1. Using independent research, write a two-page paper explaining how the eruption of Mount St. Helens in 1980 presents evidence for a global flood.

   Suggested reading material;

   Suggested websites; www.answersingenesis.org, www.icr.org, www.nwcreation.net, www.creationevangelismtools.org

Yellowstone	Section 5, Part 1 Pages 30–49	Day 95	Worksheet 3	Name

Terms to Know (Write the definition in the space below.)

thermophiles

rhyolite

supervolcano

radiometric dating

Short Answer Questions

1. One of the main geologic features of Yellowstone is the ____________ ____________.

2. What keeps the Madison River relatively warm and rich in nutrients?

3. What four types of thermal features are found at Yellowstone?

4. Explain why bison and elk living near hot springs live three to eight years less than average.

5. Why did Old Faithful's eruption pattern change in 1959?

6. When did the first eruption in Yellowstone likely occur?

Discussion Questions

1. Yellowstone was designated as our first national park in 1872. Why do you think it is important to preserve places such as Yellowstone National Park?

2. Visit www.yellowstonepark.com. See if you can find a map of the continental divide in Yellowstone National Park on this website.

Activity

1. Research the history of the Nez Perce War in 1877. Write a two-page paper explaining the conflict.

Terms to Know (Write the definition in the space below.)

quartzite

breccia

hotspot

layer

Short Answer Questions

1. What is the significance of 11,000 feet of rounded quartzite rocks found from Mount Hancock to Jackson Hole?

2. What animals are the largest members of the deer family?

3. Ice on Lake Yellowstone averages two feet thick in the winter, but is absent in spots because of _____________ _____________ on the lake bottom.

4. What could have caused such a huge deposit of volcanic debris found in the Absaroka Mountains?

5. Why is there bark missing from so many trees in Yellowstone?

6. The Bald Eagle has a wingspan reaching ______ to ______ feet.

Discussion Question

1. Have a family discussion about the wisdom of man versus the wisdom of God. Proverbs 26:12, Is. 5:21, 1 Cor. 1:18-20, and Job 38 are just a few Scripture passages that will help you get the conversation started.

Activity

1. Create a presentation comparing the marmot and the pika. How are they similar? How are they different? A great way to do this would be a PowerPoint presentation using images and facts found in the book and on the Internet. You may also create a poster using additional facts from an encyclopedia and drawing your own images!

	Yellowstone	Section 5, Part 3 Pages 66–81	Day 100	Worksheet 5	Name

Terms to Know (Write the definition in the space below.)

creationist

igneous rocks

andesite

basalt

hoodoos

erosion

Short Answer Questions

1. What causes color in the canyon walls at the Grand Canyon of the Yellowstone?

2. How did the granite boulders on Mount Washburn get there?

3. Why do the Dunraven Pass soils support a greater variety of plants?

4. How are basalt columns formed?

Discussion Questions

1. Generated by summer thunder storms, lightning is responsible for many of the wildland fires that occur throughout the western United States each year. Discuss how lodgepole pines were uniquely designed to re-vegetate an area after a forest fire.

2. Discuss the conflicting opinions and evidence presented on page 67 of the book regarding the Ice Age. What are some problems or inconsistencies with the evolutionary viewpoint of multiple ice ages forming the Grand Canyon of Yellowstone?

Activities

1. Draw a diagram of Yellowstone River Canyon that includes each of the layers described on page 79 of *Your Guide to Yellowstone*. Include a description of these layers from a creationist perspective.
2. Thomas Moran's paintings were largely responsible for the world's fascination with the western United States. His paintings provide an object lesson in the handling of paint, light, and atmosphere. You can enjoy his work by doing an Internet image search or visiting your local library. Which are your favorites?

Terms to Know (Write the definition in the space below.)

fossil

travertine

Short Answer Questions

1. Organic matter must be buried ______________ in order to be fossilized.

2. Why has the Lamar Valley been called "the American Serengeti"?

3. Why was the Lamar Valley a good place for wolf reintroduction?

4. The different colors in the hot pools and terraces ar Mammoth Hot Springs are caused by various heat-loving ________________________.

5. What enables thermophiles to live at temperatures that would kill most forms of life?

6. Geysers form when rising hot water reaches a constriction and is forced to collect. The extreme pressure keeps the water from boiling, allowing it to become superheated until it finally explodes in a display of ____________ and _____________.

Discussion Questions

1. Discuss how the Genesis Flood reasonably explains Heart Mountain Slide.

2. Compare the creationist versus the uniformitarianist explanation of the mixture of land and marine fossils found on Mount Everts.

Activity

1. Make your own geyser!

   Items needed: 12 oz. empty soda bottle, liquid soap, 2 effervescent tablets (such as alka seltzer). Be sure to do this outside or in the kitchen sink!

   Instructions: 1. Break tablets in half. 2. Fill soda bottle with very warm tap water. 3. Add a couple squirts of soap. 4. Drop tablets into bottle and immediately cover with your hand. 5. Allow the pressure to build for 3 to 5 seconds, and then remove your hand. 6. Watch your geyser erupt!

Terms to Know (Write the definition in the space below.)

metamorphic rock

brachiopod

gneiss

basement rock

moraine

dike

terrace

Short Answer Questions

1. When did homesteading start in what is now Grand Teton National Park?

2. Why do coyotes hunt cooperatively with badgers?

3. What kind of rocks make up the Teton Mountains?

4. The fur of what animal provided the financial support for early explorers of this area?

Discussion Question

1. Explain how the sedimentary layers at Moose Junction give evidence for a global flood.

Activity

1. Beavers are fascinating creatures! Prepare a presentation (poster or PowerPoint) explaining at least three fascinating facts about beavers and three ways they are beneficial to the ecosystem. Be sure to include images or your own drawings!

Terms to Know (Write the definition in the space below.)

fault scarp

glacial outwash

biodiversity

contact

Short Answer Questions

1. What sculpted the grooves in the bedrock of Cascade Canyon?

2. What do clouds, glaciers, streams, and lakes all have in common?

3. What percentage of our planet is covered with water?

4. Over half of the fresh water on earth is held in the _______ _______ and _____________.

Discussion Question

1. The water cycle is incredibly complex. Does this support your belief in a Creator? Read Job 36:27–28.

Activity

1. Create a chart or diagram illustrating how the Signal Mountain area was formed. See page 127. Use your imagination!

Terms to Know (Write the definition in the space below.)

limestone

sandstone

shale

Short Answer Questions

1. Why is biodiversity important for the health of an ecosystem?

2. What are the little semi-circular cracks on the quartzite boulders found in the terraces above the Snake River?

3. How far do pronghorn travel to avoid the winter snows?

4. ____________________ have a 320-degree field of vision.

Discussion Questions

1. Discuss how the sedimentary layers found in the Teton Range give evidence for a global flood.

2. How are pronghorn specially designed for life on the open plains?

Activity

1. "Biodiversity in a Minute" exercise. Write down the name of every plant, animal, or insect you can think of in 60 seconds. Time yourself. How many did you come up with? You may have thought of a lot – but your list will be only a small fraction of all that God created! This helps us to realize that God is bigger and more amazing than we can imagine. The Bible says, "He has made everything beautiful in its time. He has also set eternity in the hearts of men; yet they cannot fathom what God has done from beginning to end" (Ecclesiastes 3:11).

   How does understanding that the world and everything in it is special to God affect the way you care for the earth? For animals? For other people?

Terms to Know (Write the definition in the space below.)

empirical (or observable) science

historical (or forensic) science

tectonics

formation

naturalism

Short Answer Questions

1. What you believe about origins and the age of the earth depends upon your ________________.

2. Why are geology and dating methods often in the center of the creation/evolution debate?

3. The vast majority of the earth's expansive rock formations and fossil beds cannot be explained by the ____________ and _____________ __________________ we see happening today.

4. All of the world's mountain ranges show evidence of being underwater at some time in the past and many have _________________ ___________________ at or near their summits.

Discussion Questions

1. Some critics of the global Flood commonly point out that the Flood water could not cover Mount Everest. Discuss how tectonic activity in Wyoming refutes their claim.

2. Read Psalm 19:1-4, and Romans 1:20. Discuss some ways you have personally observed the hand of a Creator in nature.

Activity

1. Find a beautiful and quiet place outdoors – in your own yard or a park. Observe the sights, sounds, and smells of nature. Write your own psalm of praise to God for His magnificent creation! Read Psalm 104 for an example, but make your psalm as long or as short as you wish.

Terms to Know (Write the definition in the space below.)

Channeled Scabland

carbon dating

geologic column

parent element

daughter element

paleontologist

Short Answer Questions

1. Yellowstone and the Tetons cover a massive amount of territory and present very different landscapes. Yellowstone is largely rolling hills and valleys carved by the movement of _________, while the Tetons are mostly the result of ____________ _____________.

2. The horizontal boundary lines between sedimentary rock layers are called _____________.

3. Moraines are formed as a glacier moves, accumulating debris. Debris built up and left in front of a glacier is called a _______________ moraine, while the debris left on the sides are ________________ moraines.

4. The measurement of radioactive elements as they decay from a parent element to its daughter element, which occurs at what is thought to be a know rate, is a process called __________________ _________.

Discussion Question

1. There are many methods used to date rocks. Your book gives a brief discussion of three: carbon dating, radiometric dating, and sedimentary dating. Discuss why it is important to understand the basic assumptions being made relative to these methods. For a more in-depth discussion of these methods and assumptions see:
http://www.answersingenesis.org/articles/ee2/dating-methods

	Yellowstone	Sections 9–10 Pages 158–171	Day 118	Worksheet 12	Name

Terms to Know (Write the definition in the space below.)

transitional fossil

biotic communities

mutation

theory

Short Answer Questions

1. The conditions necessary to make a fossil are very uncommon. First, the organism must be ______________ ______________ to protect it from decay and scavengers.

2. The second condition necessary to make fossils is that ________________ must be absorbed rapidly by the organism before it decays.

3. By far, the most common tree in Grand Teton and Yellowstone is the ______________ __________.

4. Next to the moose, _________ are the largest member of the deer family.

5. Wolves are big. They can be _________ feet long from the tip of their nose to the tip of their tail!

6. A major cause of the decline of the American Plains Indians can be attributed to the loss of _____________ as their primary food source.

Discussion Questions

1. Why would Noah's Flood have been an ideal environment for fossils to form?

2. Discuss how the fire suppression programs implemented in Yellowstone around a century ago contributed to the fire in 1988. Did man unknowingly interfere with God's perfect design for forest management? What was the result?

Glacier National Park Worksheets

for Use with

Explore Glacier National Park with Noah Justice
(Awesome Science Series)

Chapters 1–2: Opening and History of the Park
About Glacier National Park

Fill in the blanks with words from the following list:

Canada	Cheyenne	Crown of the Continent	Flathead	genealogies	global Flood
ice ages	Lake McDonald	Montana	pollution	President Taft	
proterozoic	Salish	Shoshone	stomatolites		

1. Glacier National Park sits on the border of _______________ and _______________.

2. Secular scientists say these rocks are over 1.6 billion years old and contain _______________ built by microbes supposed to have formed over a billion years ago, according to evolution.

3. They also say many _______________ over millions of years carved these valleys.

4. The Bible indicates these rocks were formed quickly during the _______________, which also produced the right conditions for one Ice Age, all happening within the last 4,350 years. This time scale does not match the billions of years proposed by secular scientists.

5. The history of this area goes back thousands of years. The earliest inhabitants were American Indians: the _______________, _______________, _______________, and _______________.

6. The anthropologist, historian, and writer George Bird Grinnell called this the _______________. He made it his life goal to preserve this beauty for all mankind.

7. In concern that the land not be exploited for private enterprise, _______________ in 1910 established this area as a national park.

8. The park encompasses over 1.2 million acres, with two mountain ranges, and over 700 pristine lakes. The deepest is _______________at 464 feet deep. The park has over 200 waterfalls, some almost 500 feet tall.

9. Glacier National Park is part of a large preserved ecosystem known as the "Crown of the Continent Ecosystem." There is little to no _______________, since there are no major dense human population centers or factories.

10. The rocks in Glacier are supposed to be the best-preserved _______________ sedimentary layers in the world, showcasing the earth's most fruitful source for the record of "early life" according to evolutionary thought.

11. But the Bible indicates the earth was created about six thousand years ago using the _______________ found in Genesis and Matthew.

Chapter 3: Lewis Overthrust
Secular Ideas for Mountain Formation

Please note whether the following statements are true (T) or false (F):

1. Glacier National Park is in the Smoky Mountains. ________
2. Secular geologists believe the Rocky Mountains formed over 1.6 billion to 800 million years ago. ________
3. When one rock layer is pushed over another through tectonic activity, scientists call it an overthrust. ________
4. We can see overthrusts happen today and perform them in experiments. ________
5. All rock layers in the Rocky Mountains show evidence of being pushed in the same direction. ________

Chapters 6–7: Ice Age Introduction and Ice Age
The Ice Age

Fill in the blanks with words from the following list:

abundant moisture	aerosols	ash	computer model	cool summers
mild winters moraines	observe	oceans	snowball earth	sub-zero
uniformitarianism	U-shaped valleys			

1. ________________ are formed when an ice sheet moves down a valley, eroding the valley walls and valley bottom.

2. ________________ are caused by a moving glacier. As the ice moves down a valley, the weight of the glacier pushes rocks and debris to the side or out in front of it.

3. When secular scientists first discovered evidence for an Ice Age, it didn't fit their ________________ view, which says that the processes we see today are the ones that happened in the past.

4. Secular scientists claim we have seen at least four major Ice Ages, and some scientists even suggest our distant past included ice everywhere, calling it ________________.

5. The primary elements needed for an ice age are ________________, ________________, and ________________.

6. Abundant moisture usually comes from warm ____________. Most of our large storms and hurricanes come from the equatorial regions where the water is extremely warm. But to develop the vast amounts of moisture to create an ice age would require a much warmer ocean worldwide.

7. Next, you need mild winters. If winters are too cold, the air cannot hold moisture, and therefore it cannot snow. We see this in the Arctic, where temperatures get into the ____________ level and snowfall is absent.

8. Finally, you need cool summers. Mechanisms such as volcanoes can put ____________ and ____________ into the atmosphere and cause global cooling for short periods of time. This cooling usually lasts just a couple of years for one major event.

9. In the uniformitarian model, it would be very difficult to generate these conditions at the same time for just one ice age, not to mention many ice ages. Every ________________ they have run does not work. It puts ice in the wrong places, cools down the wrong places, or doesn't produce the right amounts of ice.

10. Since no one was there to ________________ and record these events happening, their ideas are just presumptions.

The Biblical Answer

Please note whether the following statements are true (T) or false (F):

1. Sedimentary rocks are formed by water and under water. ________
2. The Bible says God decided to destroy everything on earth by water millions of years ago. ________
3. The global Flood lasted for 150 days. ________
4. Sediment layers from the beginning of the Flood contain the earliest buried animal fossils. ________

Chapter 8: Biblical Ice Age
The Biblical Ice Age — Massive Storms and Snows

Fill in the blanks with words from the following list:

creationists catastrophic friction precipitation ice sheets meteorological

1. As Bible-believing ____________________, we have a good answer for how the ice age formed.

2. During the global Flood, the earth was going through ____________________ geologic changes. At the beginning of the Flood, the Bible says the fountains of the great deep burst forth. This included warm subterranean water from within the earth's crust, which was ejected onto the surface. In addition, the fountains of the great deep included molten lava.

3. Large-scale catastrophic tectonic activity would have caused ____________________. The water on top of the tectonic plates would have heated dramatically.

4. These conditions would have caused tremendous evaporation, resulting in very large ________________ near the shores, and snowstorms in the higher latitudes.

5. Most of the mountain chains we see today show evidence of them rising quickly during and after the Flood. The massive storms moving across these mountains would have dumped large amounts of snow in these regions, and the snow packs would eventually turn into ____________________.

6. This snow and ice would have eventually filled in the valleys and created large ice sheets over the northern and southern continents. With the ____________________ conditions right after the Flood, building the Ice Age quickly was not a problem.

Massive Melting

Discussion Questions

1. What happened during the Missoula Flood in the Pacific Northwest?

2. How do we know large regional floods, like the Missoula Flood, occurred?

3. What are erratic boulders?

Chapter 10: Another Ice Age?

Fill in the blanks with words from the following list:

promise reoccurring scientific basis

1. Secular scientists say that ice ages are a ____________________ event.
2. If man ignores the biblical Flood, then he loses all hope for knowing what caused the Ice Age. If they think they just happen over and over with no real ____________________ for how it all began, they are just speculating that it will happen again.
3. We know it won't happen because there will never be another global Flood. This was God's ____________________. What the Flood and subsequent Ice Ages shows us is that God is serious about His judgments.

Chapter 11: Global Warming

Fill in the blanks with words from the following list:

active carbon dioxide diatoms global warming Little Ice Age melting

ozone layer stewards

1. One hundred years ago, there were 150 known glaciers in the park. Now, only 37 remain, and only 25 are considered ________________ glaciers.
2. An active glacier is at least 25 acres in size. This decrease in glaciers is seen worldwide and has accelerated since 1980. Based on current __________________ rates, it is estimated that all glaciers in the park will disappear by 2020.
3. Some scientists blame man for ____________________ and the melting of glaciers, but there are many more powerful mechanisms that could cause our world's climate to change.
4. In the 1600–1800s, the world went through the ____________________, where glaciers advanced and summer crops failed. We now know this was caused by sunspots and volcanism, not man.
5. Forests around the world are now growing at amazing rates because they absorb __________________ and other man-made gasses.

6. The oceans are full of ________________, which also take in the bad gasses and give back good ones.

7. Our atmosphere's ________________ isn't being impacted by man-made gasses either. The growth of giant holes in the ozone layer has been proveen to be caused by other natural factors.

8. It's true that God put man in charge of the earth and to take care of it. We should be wise ______________, but there is a misguided campaign to say that man is destroying the earth.

Dinosaur National Monument Worksheets

for Use with

Explore Dinosaur National Monument with Noah Justice
(Awesome Science Series)

Chapter 2: History Features, Tour Monument, and Ancient Indians Dinosaur National Monument

Fill in the blanks with words from the following list:

Allosaurus Camarasaurus fossilized evolutionary Freemont Morrison Foundation

paleontologist petroglyphs sedimentary Tower of Babel

Note: One of these concepts will be used twice.

1. Near Dinosaur, Colorado, in 1909, a large deposit of dinosaur bones was found by Earl Douglass, a ________________________, working and collecting for the Carnegie Museum of Natural History.

2. The most common dinosaurs found here are the ____________________ and ____________________.

3. They are found in ____________________ rock layers of sandstone, mudstone, siltstone, limestone, and a conglomerate bed, the layers that locally make up the Morrison Formation.

4. The ____________________ is found in the western United States covering some 1.5 million square miles. Only a tiny fraction is exposed and accessible to geologists and paleontologists.

5. The Monument is also known for its plentiful Native American ____________________, many of which can be explored along the access road, near the quarry.

6. Dinosaur Quarry is the most famous sight of the monument. It is enclosed, taking you right up to the steeply tilted rocky layers, which contain hundreds of ____________________ dinosaur bones. This large quarry building keeps the fossils safe from erosion and vandalism.

7. Many of the displays reflect an ____________________ worldview, often portraying millions of years as fact.

8. Petroglyph drawings were made by the ____________________ people, a tribe who lived in the area from A.D. 200 to 1,300. They lived in small bands or family groups, grew crops, and didn't build large permanent dwellings, like their neighbors at Mesa Verde.

9. ____________________ are patterns, chipped or carved into the rock, while pictographs are patterns painted on the rock. Pictographs are relatively rare here, because they are more easily weathered.

10. After Noah and his family came off the Ark, their offspring populated the Middle East. After the ____________________, large groups of people moved away from there, to populate the earth. The Freemont people may have been some of the first settlers of this area.

Bonus Activity

Using a map, atlas, or the Internet, locate the Dinosaur National Monument. How close do you live to the park?

Chapter 5: Two Worldviews

Fill in the blanks with words from the following list:

biblical billions corrupted Creator dinosaurs evolved law-giver

naturalism salvation scientific laws special creation

1. As you tour the monument, it's important to realize there are two worldviews to be considered: ________________ and ________________.

2. Naturalism says that over ______________________ of years, the universe and earth came about by chance, through slow and gradual processes.

3. A key part of this worldview is that life ______________________ from simple to complex forms, through mutations and natural selection; a process called biological evolution.

4. ____________________ are just a part of the evolutionary history. According to some evolutionists, they were here for a few hundred million years, then died off quickly. Others believe they are alive today as birds.

5. ____________________ are a description of how God created the universe to operate and how He upholds His creation. We can demonstrate these laws by repeating experiments and observations, thus showing that the results are always the same under given conditions. They are set in place and only God can operate outside of these laws.

6. From what we observe, the __________________ worldview makes much more sense in nature. We remark on the beauty and originality of each kind of animal, perceiving incredible design. Scientific laws are put here by our Creator to provide consistency and dependability.

7. The second law of thermodynamics establishes that everything goes from order to disorder, from usable to unusable. We know that matter plus time plus chance does not equal great designs, but causes our world to fall apart. For our world to keep going, it needs a ______________ and __________________ to hold things together.

8. The Bible reveals who this creator is. He's the God of the Bible. The Bible says He created a perfect world, in six literal days, about six thousand years ago. The first man, Adam, sinned and brought a curse on creation, and sin __________________ all of humanity.

9. The earth went through a worldwide judgment in the global Flood. God provides __________________ for mankind through His Son, Jesus Christ, and one day He will create a new heaven and earth.

Bonus Activity:

Read how God created the earth and how humans allowed sin to enter the world in Genesis 1–3.

Chapter 7: Finding Dinosaurs

Fill in the blanks with words from the following list:

Flood iridium K-T Boundary quick burial sedimentary volcanoes

Note: One of these concepts will be used twice.

1. All dinosaurs are found in ____________________ layers of sand, silt, and mud, which would seem to indicate they were buried quickly in a flood.

2. We can assume this because any dead animal left in the open air would quickly decompose and scavengers would eat the remains. Only ____________________ would keep the animal intact and provide the right conditions for the bones to be quickly fossilized.

3. All dinosaur fossils are below what is called the ____________________ layer, which is a distinctive layer of sediment in between other rock layers, and is thought by evolutionists to represent a mass extinction event. This is because many species are found below this layer and not above it.

4. They believe this because the layer contains ________________, an element found in meteorites and some impact sites, such as Meteor Crater in Arizona, and not in most earth rocks.

5. There are two reasons why ________________ in a meteorite from an asteroid does not make sense as the sole source for the iridium found in this layer. The dust from an asteroid does not spread very far, and the amount found in this layer in various places around the earth is just too much for a single impact from an asteroid fragment to produce.

6. Iridium is also found in ash from ____________________. With the evidence of many super-volcanic eruptions around the world in the past, they could easily have produced the amount of iridium found in this layer around the earth.

7. There is a lot of evidence to show there was massive volcanic activity during and after the ____________________. Large lava flows and ash deposits were laid down between the sediment layers across the earth's surface during this worldwide catastrophe, and as the Flood waned and the earth was settling to today's relative equilibrium.

As the Flood waters then receded off the earth, more sediments would have be laid on top of the K-T Boundary layer, burying the dinosaurs even farther. In addition, the surface of the earth, below the waters, would have been rising over vast areas and the receding waters would have eroded vast areas and carved valleys and canyons.

Chapter 8: The Allosaurus
Dinosaurs at the Monument

Fill in the blanks with words from the following list:

bipedal evolutionists Fall plants predators different lizard

1. One of the main dinosaurs found at the monument is the Allosaurus, which was supposed to have lived between 155 to 150 million years ago, according to ________________.

2. Allosaurus means ______________________ and they are found in the Morrison Formation, mainly here in Utah and Colorado, but also in Wyoming, Montana, South Dakota, Oklahoma, and New Mexico, as well as in other layers in Portugal and possibly Tanzania.

3. It is a large __________________________ dinosaur, meaning it walked on only two feet. Its skull was large and equipped with dozens of large, sharp teeth. It averaged about 28 feet in length, with some reaching almost 40 feet.

4. Many secular paleontologists think dinosaurs like the Allosaurus were ______________________. The fact is, we just don't know. It's a big assumption that just because an animal has sharp teeth it was a predator.

5. Genesis 1:29–31 tells us that before the Fall in the Garden of Eden, man and animals were commanded by God to eat from ____________________. There was no animal death, prior to the Fall — they were NOT meat-eaters.

6. The fact is, we just don't know what some dinosaurs ate after the ______________________. It's terrifying to think of packs of dinosaurs hunting down their prey, but this is a story with little evidence to support it. There's evidence in the fossil record to suggest some dinosaurs were scavengers and hunters, but not all of them were. Even the intimidating Allosaurus could have been a vegetarian.

Chapter 9: Death of Dinosaurs
What Happened to the Dinosaurs?

Fill in the blanks with words from the following list:

fossil record fossilized habitat Marco Polo minerals quick burial rebellion

simultaneous small horse variety violent

1. We know about dinosaurs because of their ____________________ bones in sedimentary layers. Many have been found in an opisthotonic pose, where the head is thrown backward, the neck and body arched, and the tail thrown backward. No one knows exactly why this death pose is so frequent.

2. Dinosaur skeletons around the world seem to reveal a ________________ death during water transport. These clues give us a good idea as to how these dinosaurs died off so suddenly. Those who believe in evolution have a myriad of ideas about the death of the dinosaurs.

3. One big clue to their demise is the way in which they are buried. Their bones are spread out across vast areas in sedimentary layers deposited by water. Rarely are they found as complete skeletons. Most of the bones have fossilized, which is a rare process and only happens when an animal is quickly buried and the right ____________________________ can replace the bony tissues so the bones turn to stone.

4. Starting with the creation account found in the Bible, God created all the land animals on day 6, which would have included dinosaurs. For about 1,600 years dinosaurs grew in number and ____________________. The Bible and the fossil record seem to indicate the earth was much more lush and temperate; a perfect place for dinosaurs to grow big and plentiful.

5. When God came to Noah about 1,600 years after the creation of the world, He told Noah he was going to flood the entire earth because of man's ________________. He told Noah to construct an ark and take two of every kind of land animal and seven of some onto the ark, which would have included pairs of dinosaurs.

6. Noah didn't have to take large, older adult dinosaurs onto the ark, but smaller, young adults would have much more life left to reproduce young, once the Flood receded. When we think of dinosaurs, we often think of huge creatures, but the average size of a dinosaur was only about the size of a ____________________.

7. There are about 670 dinosaurs named from the ____________________, but there are only about 55 kinds of dinosaurs, so this is about how many pairs God sent to be taken on the ark. With over 2.2 million cubic feet of space within the ark, carrying the dinosaurs and all the other kinds of land animals wouldn't have been a problem. There would have been more than enough space for the animals, food, and Noah's family.

8. Once the Flood ended and the animals came off the ark, the earth's ________________ was much different than before the Flood. As the dinosaurs spread out, it's likely they didn't repopulate in the same numbers as before the Flood because their environment had drastically changed. Food sources and the different climate would have drastically limited their repopulation.

9. With the vast graveyards of dinosaur bones we find around the world, the demise of the dinosaurs would have to have been catastrophic, a nearly ______________________ global destruction.

10. In addition, many of the intact skeletons we find show signs of drowning, then ______________________ in mineral-rich waters with a lot of silt. This is what we would expect to find if the Flood were true. As the fountains of the great deep came up out of the earth, these fountains would have included warm mineral-rich water mixed with volcanic ash and sediments. All of these elements would have created the right conditions to quickly fossilize the bones.

11. There are different accounts in human history where strange dragon-like creatures were described in literature. People like ______________________ and the Romans documented encounters with these creatures. Their descriptions sure do sound a lot like dinosaurs.

Chapter 10: Dinosaur Soft Tissue

Please note if the following statements are true (T) or false (F):

1. _____ Mary Schweitzer discovered soft tissue in a T. rex bone in the 1990s.
2. _____ Most secular scientists believe dinosaurs died about 65 million years ago.
3. _____ Soft tissue lasts for millions of years.
4. _____ A biblical worldview holds that the earth is millions of years old.
5. _____ Creation scientist Mark Armitage discovered that soft dinosaur tissue might be abundant.
6. _____ Paleontologists in China have found soft tissue, which supports Mark's claim.
7. _____ Iron does not preserve soft tissue.

Dinosaurs and Humans

Fill in the blanks with words from the following list:

behemoth Congo dragons Job leviathan Peru Turkmenistan

Angkor, Cambodia

1. In the temple ruins at ____________________, there is a passageway with ancient artwork in the stones. The artwork depicts many animals found in the area such as elephants, fish, snakes, and monkeys. There are no mythological figures among the architecture, so it appears they made carvings of animals they saw at the time. But an unusual animal stands out among the rest — it's a stegosaurs-like figure in the stone. This temple was erected for a king about a thousand years ago. As far as we know, this ancient culture didn't study paleontology.

2. In __________________, clay jars, textiles, and stones have been decorated with dinosaur-like creatures from cultures that pre-date the Incas. Secular scientists describe them as mythical creatures, but they clearly resemble various dinosaur kinds.

3. Then there is a site in __________________, which borders Iran, where what appear to be human footprints lie alongside thousands of dinosaur prints.

4. In the ________________, it is reported that a dinosaur-like creature may be living where men hardly visit. The natives in the deep Congo jungle have claimed to have seen this mysterious creature and call it Mokele Mbembe. When these natives were asked to identify this creature, they were given drawings of different jungle creatures as well as a drawing of a dinosaur. They quickly identified the creature as a dinosaur.

5. One of the most ancient books of the Bible, the Book of ________________ mentions several curious creatures described in detail, seeming to resemble a dinosaur and a large sea creature.

6. In Job 41, God describes a large land animal called ____________________, with bones as tough as iron and a tail as big as a cedar tree.

7. God describes to Job another creature living in the sea — __________________. It cannot be captured and has a double coat of armor. Its teeth are fearsome and it breathes fire. Some have suggested this creature was a large crocodile. But the largest croc ever found doesn't match the description.

8. Before the word dinosaur was invented, these types of creatures used to be called _______________, and the Bible uses this word quite a bit in the Old Testament. In fact, if you did a word search on "dragon" in the King James Version, there are over 30 references in the Old Testament, and some of these references surely refer to dinosaurs or large sea creatures like plesiosaurs.

The Mammoth Site Worksheets

for Use with

Explore the Mammoth Site with Noah Justice
(Awesome Science Series)

Chapters 1–3: Opening, History Features, and Secular Theories About the Mammoth Site

Fill in the blanks with words from the following list:

Bering Strait	global Flood	mineral water	secular	sinkhole
underground springs	mammal fossils	worldview	woolly mammoths	Colombian
mammoths	flora fossil evidence	geologic	Colombian mammoths	

Note: One of these concepts will be used twice.

1. As they examined the bones, they found mainly __________________, but also a few __________________.

2. They have found a number of other __________________, some of which are extinct.

3. Hot Springs is known for its ____________________ and ____________________.

4. It is thought that this area was once a ________________ which mammoths and other mammals fell into and couldn't get out of, so they died here.

5. Because of __________________, it's been suggested this area was once a cold, treeless grassland with shrubby bush, water-rooted plants, sedges, and grasses.

6. __________________ scientists think this all happened about 26,000 years ago. They assume this long age because of their ______________________.

7. The Bible indicates that about 4,350 years ago, the __________________ destroyed the entire earth.

8. With the catastrophic ____________________ activities going on during the Flood, and their waning thereafter, the oceans heated up, providing perfect conditions for increased evaporation that resulted in a massive amount of snow and ice accumulating in the mountains, so that 30 percent of the continents were eventually covered with ice in less than 300 years.

9. The ____________________ area between modern-day Russia and Alaska was dry land, and animals and man could walk across.

10. Several thousand years ago, this area of South Dakota was near the southern edge of the ice sheet where mostly ____________________ roamed.

Discussion Questions:

1. Why did Phil Anderson decide to sell the land where the Mammoth Site is?

2. Why did scientists mostly find young adult male mammoths at the site?

3. How does a Christian worldview explain that the Mammoth Site events did not happen 26,000 years ago?

Bonus Activity:

Read what the Bible says about the Flood in Genesis 6–8.

Chapter 6: Carbon Dating

Fill in the blanks with words from the following list:

assumptions	atmosphere	Bible	biosphere	carbon 14	contamination	Flood
half-life	magnetic field	radiocarbon	sediments	specimen		

1. ____________________ dating was used by secular scientists here at the site to conclude these fossils are about 26,000 years old.

2. Since the Bible indicates the ____________________ happened about 4,350 years ago, the events here at the Mammoth Site had to have happened less than 4,350 years ago.

3. The very fact that these fossils are near the surface tells us that they died after the Flood, because the Flood laid down thousands of feet of ____________________ below these mammoth fossils.

4. The time it takes for half of the radiocarbon to decay is called its ____________________.

5. There are some basic ____________________ made by these scientists when radiocarbon dating is used.

6. First, they assume ____________________ has always been produced at the same rate as it is being produced now.

7. Second, they assume there was no ____________________ of the sample by outside sources of carbon 14.

8. Third, they assume they know the amount of carbon 14 the ____________________ started with.

9. Fourth, they assume the ____________________ has had the same carbon 14 concentration in the past as it does now.

10. Fifth, the ____________________, or place on earth where organisms live, has always had the same overall carbon 14 concentration as the atmosphere.

11. With the ____________________ as our guide on the age of the earth, we can also use other scientific investigations to show radiocarbon dating may not be the most reliable method for dating these fossils.

12. In the case of the mammoths found at Hot Springs, because they lived after the Flood, the 26,000-year age is inaccurate, mainly due to the change in the earth's ____________________ since the Flood and the inaccurate assumptions used by the secular scientists.

Discussion Questions:

1. How does carbon 14 enter the animals' bodies before they become fossils?

2. Why is it impossible to use radiocarbon dating for things that are supposed to be millions of years old?

3. Why do the assumptions made by secular scientists make the secular dates for the Mammoth Site unreliable?

Chapter 7: What Is a Mammoth?
More about Mammoths

Please note if the following statements are true (T) or false (F):

1. _____ God made land-dwelling animals on day two.
2. _____ Mammoths belong to the scientific order called proboscideans, which means an animal with a long, flexible snout or trunk.
3. _____ Noah took only one male and one female of the elephant kind on the ark, so elephants, mastodons, and mammoths all come from a common ancestor.
4. _____ Elephant kinds living in hot areas had long hair, but those living in cold areas had little hair.
5. _____ God created the elephant kind 70 million years ago.
6. _____ Mammoths had teeth with high pointed ridges.

Discussion Questions:

1. How many mammoths do scientists estimate are in the sinkhole?

2. What was the mammoths' main predator?

3. How often do modern-day elephants produce one offspring?

4. Why were mammoth populations able to grow so quickly?

Chapter 10: The Ice Age and the Bible

Fill in the blanks with words from the following list:

archeological conditions evaporation fossil and historical record ice sheets melted volcanic

1. With warm oceans worldwide at the end of the Flood, ______________________ was intense, creating very large storms across the continents. This fell as snow in the highlands.

2. The earth was also settling down geologically for a hundred years or so after the Flood, with large amounts of ______________________ activity. This activity put huge amounts of ash and aerosols into the atmosphere, keeping the summers cool by reflecting sunlight back into outer space.

3. The warm oceans, warm winters, and ash in the atmosphere causing cool summers led to a quick accumulation of the ______________________, which we now call the Ice Age.

4. Only the__________________ provided by the Flood can produce the Ice Age.

5. Scientists who believe in the Bible's timeline of history have tried to calculate and model how quickly an Ice Age could develop. They believe that it could take only 500–700 years. But biblical time constraints and ______________________ evidence would also limit the Ice Age to only about 300 years.

6. There is also strong evidence this Ice Age __________________ catastrophically within about 30–70 years.

7. This rapid build-up and melting of the ice would have had tremendous impacts on the climate, the animal population, and humans. And this is exactly what we see in the ___________________________.

Discussion Question:

Why do secular scientists struggle to find answers to problems like extinction?

Chapter 10: Mammoth Extinctions
Secular Ideas of Extinctions

Fill in the blanks with words from the following list:

asteroid   climate change   comet   extinctions   humans   theory   younger

1. ________________ do happen. But this event of massive mammal extinctions around the same time is unique, and indications are it all happened near the end of the Ice Age.

2. Some think ____________________ may have driven groups of mammoths and other large mammals over cliffs and then slaughtered and butchered them. There are a couple of problems with this idea.

3. A second idea is that hunters may have selected _____________________ mammoths as their prey. The youth were considered slower and less experienced at evading hunters. The young were being killed faster than they could be replaced.

4. Another __________________ says that hunters killed the mammoths in a swift and massive attack before the mammoths learned to avoid human hunters. But there were too many mammoths for this to happen and not enough humans.

5. Some have even suggested a _________________ or ________________, but there is no evidence for this either.

6. More recently, secular scientists have suggested ______________________ played a key role in their demise, and this is pretty close to the truth.

Chapter 11: Biblical Answer
The Ice Age and the Bible

Fill in the blanks with words from the following list:

climate Ice Age permafrost silt temperatures water table windstorms

1. A big clue to how the mammoths died can be found in how they were buried. Most mammoths we find today are found in layers of silt and ________________________, which is mostly frozen, wind-blown silt. Where did this silt come from?

2. As the glaciers grew and made their ways into the valleys, the ________________________ changed. Eventually the oceans cooled and the snowstorms became less dramatic, and there was also a decrease in volcanic activity. Instead of the ice sheets building, they began to melt.

3. As the ice retreated, vast amounts of glacial melt water entered the oceans. At first this fresh cold water stayed on the surface, covering the warm oceans, ________________________ dropped quickly and eventually there was a layer of ice on the seas.

As the ice retreated, ground-up soil from under the ice was very fine and was easily transported by wind. As the seas cooled in the north, winters became cooler and the summers began to heat up worldwide.

4. This huge change in hot to cold climates created massive ________________________, which picked up much of the silt and created immense dust storms.

As herds of mammoths encountered these storms, they would aim their backsides to the wind, hoping it would pass. But just like a snowdrift, the wind-blown silt began to pile up against their bodies.

5. Before long, the mammoth would be completely covered in ________________________. They would inhale the silt and suffocate. If they tried to wiggle out, the packed silt would act like packed snow. It's very difficult to get out of. They could end up breaking their leg bones just by the exerted force of trying to get free.

6. The ________________________ from below would quickly come up into the wind-blown silt. In the northern parts of the continent, the water would freeze and the ground would become permafrost quickly.

Larger mammals would also be wiped out in storms like this. This idea of large dust storms could easily explain the mass extinction at the end of the Ice Age. When we go back and examine the Mammoth Site, the sinkhole has been filled by silt. Some mammoths were buried quickly in this silt. When these dust storms came, the dust would have settled in the sinkhole, quickly burying the carcasses. This would also explain why so many bones have been preserved.

7. The secular scientists claim the process of animals being trapped and windblown silt covering their remains would have happened in about 300–700 years. However, the rapid end to the post-Flood ________________________ accelerated this process into about 50–100 years of time, burying many mammoths very quickly.

Arches National Park and Natural Bridges National Monument Worksheets

for Use with

Explore Arches National Park and Natural Bridges National Monument with Noah Justice
(Awesome Science Series)

Chapter 2: Arches Introduction
About Arches National Park

Fill in the blanks with words from the following list:

10 43 2,000 4,000 Colorado erosion slickrock

1. In the southwest part of the United States, on the border of Utah and ______________, is one of the most unique national parks in the country.

2. Sitting on almost 80,000 acres, it is the largest concentration of natural arches in the country, containing more than ____________, of these fascinating geologic formations.

3. Over ________________ arches are known to have collapsed since 1977, so the number of arches is decreasing.

4. In addition to the arches are incredible balanced rocks, spires, pinnacles, and ____________ domes.

5. The park sits between __________ and 5,600 feet above sea level, near Moab, Utah.

6. The arches are thought to be the result of water ____________, but the nearest sea is over 700 miles away and about a mile lower in elevation.

7. The park receives an average of _______ inches of rain a year, but scientific research has shown this area received a lot more precipitation in the past.

Chapter 2: Arches Introduction
History of Arches

Fill in the blanks with words from the following list:

80,000 Anasazi educational Ice Age monument petroglyphs pictographs tourist

1. The rich history of petroglyphs and pictographs on rocks in the park is evidence that humans lived here since the ________ __________ about 4,000 years ago.

2. ________________ are motifs that are pecked, ground, incised, abraded, or scratched on the rock surface.

3. _________________ are paintings or drawings in one or more colors, using mineral pigments and plant dyes on the rock surface.

4. From the discovery of ancient dwellings in the park, it's thought that the Freemont and _______________ people lived here as recently as 700 years ago.

5. The Arches area became more well-known when a railroad manager and a photographer visited the area in 1923 to scout it for a possible _______________ destination.

6. Hebert Hoover signed the proclamation making the area a national _____________.

7. The idea behind the national monument, was to protect the arches, spires, and other formations for scientific and _______________ value.

8. After various presidents made proclamations to enlarge the park, it was expanded to almost ________________ acres.

Chapter 3: What Is an Arch?

Please note whether the following statements are true (T) or false (F):

1. _____ An arch opening must measure at least three feet across, but the height can vary.
2. _____ The arch can stand independently or be a part of a rock system.
3. _____ The largest arch in the park is Landscape Arch, spanning over 306 feet. It is the world's largest arch.
4. _____ According to naturalists, new arches are constantly forming, and these formations are easily observed.
5. _____ The five observable attributes used in the classification of arches are contextual, morphologic, metrics of the arch, geologic observations, and anthropomorphic classification.

Chapter 4: What to See
What to See at Arches

Fill in the blanks with words from the following list:

Babel fiery Moab peregrine trails

1. Courthouse Towers includes Sheep Rock, Three Gossips, the Organ, and the Tower of ________.
2. When coming to Arches, the main entrance is on the south side of the park near __________, Utah.
3. Over the years, water and wind erosion have worn down the sandstone into a huge collection of spires and fins on the formation called ________ Furnace.
4. The park draws many hikers, mountain-bikers, and off-road enthusiasts each year for miles of ____________, with incredible views.
5. Keep your eyes out for wildlife such as spade-foot toads, antelope, __________ falcons, desert big-horn sheep, mule deer, midget faded rattlesnake, and if you're blessed, a mountain lion.

Chapters 5–6: The Secular View and the Flood is the Cause
How Did It Happen?

Fill in the blanks with words from the following list:

150 300 4,000 70,000 catastrophic days erosion quick

1. Secular scientists claim that _____ million years ago an inland sea covered this area, leaving behind salt beds that are thousands of feet thick.
2. Secular scientists say that _______ million years ago the soft, red sandstone was deposited.
3. Secular geologists have estimated it took about ___________ years of water, frost, and wind operating in a dry climate to form isolated Delicate Arch.
4. If we use current _________ rates, these arches should have weathered and collapsed long ago.

5. These amazingly beautiful, freestanding arches seem to be a relic of past geological processes that formed them by some mechanism using ________ erosion.

6. During the Flood, hot salty volcanic waters gushed from inside the earth, and on contact with the cooler surging Flood waters, massive layers of salt and sandstone were laid down in a matter of ________.

7. The fact that bridges and arches are not forming today, but have at a high rate in the past, would give good motivation for believing in ____________ processes, such as a global Flood.

8. Because of natural erosion, over the last ______ years, the arches would have naturally eroded away and become smaller, leaving the 2,000 or so arches we find today.

Chapter 7: Natural Bridges Introduction
About Natural Bridges

Fill in the blanks with words from the following list:

erosion dinosaur Hopi river White

1. In the southeast corner of Utah, water erosion has made some fascinating features in _______ Canyon, also known as Natural Bridges National Monument.
2. Even though they are different formations from the Arches, it is believed that water __________ was also how they were formed.
3. The three bridges are the Sipapu, Kachina, and Owachomo, all of which are _______ Indian names.
4. Inside the monument, ancient Indians left evidence for their residence, including an etching that looks a lot like a sauropod ___________.
5. There is no permanent ________here to carve this canyon, only periodic flash floods.

History of Natural Bridges

Please note if the following statements are true (T) or false (F):

1. ____ The earliest documented exploration of this canyon happened in 1893 by *National Geographic Magazine.*
2. ____ President Theodore Roosevelt designated it as a national monument, which was actually Utah's first national monument.
3. ____ It is thought by anthropologists that the Anasazi Indians occupied this area between 500 B.C. and A.D. 1270.
4. ____ The earliest inhabitants probably lived in pit houses on the mesa tops, while the later Anasazi built cliff dwellings that can still be seen today.
5. ____ Though once accessible only by horseback, in the 1950s the gold boom resulted in the creation of new roads into the area and eventually a state highway was paved in 1976.

Chapter 7: Natural Bridges Introduction
What to See at Natural Bridges

Fill in the blanks with words from the following list:

Hopi Kachina Owachomo Park Sipapu trails

1. After you pass the visitor's center on the east side of the park, the first natural bridge you come to is __________ Bridge, the fourth largest in the world.

2. There are _________ that allow you to hike underneath each bridge, and these are rated medium to difficult, since many have ladders, stairs, switchbacks, and short, steep sections of slick rock.

3. ___________ natural bridge is larger than Owachomo but smaller than Sipapu, and in June 1992, approximately 4,000 tons of sandstone fell from the inside of the this bridge.

4. Government surveyor William Douglas assumed the ancestral Puebloan people who once lived here were related to the present-day _________ people.

5. The smallest and thinnest of the three natural bridges is called ________________.

6. The International Dark Sky Association named Natural Bridges the world's first "International Dark Sky _________."

Chapter 8: The Geology
Geology of the Southwest and Natural Bridges

Fill in the blanks with words from the following list:

catastrophic	cavitation	days	Flood	historical	lower	sandstone
seas	sedimentary	surface	turbulent	water		

1. Much of this area in Utah is made up of sedimentary layers of ____________.

2. Just to the southwest of Natural Bridges is the Grand Canyon, where some catastrophic event eroded over 5,000 feet of ________________ rock layers.

3. Secular geologists believe inland _________ and deserts, over millions of years, laid down these layers.

4. The evidence is much more consistent with the layers being all laid down very quickly by _________.

5. Cross beds can be formed by desert wind or moving water, determined by the downslope of the angles of the dunes, with ________ angles indicating they were formed under water.

6. Because fossils are rare in the sandstones in the area of Natural Bridges, it is classified as a high-energy environment, meaning that grains of sand were deposited in highly ____________waters.

7. Forensic science investigates all possible elements of a past process, trying to come up with an idea of how it happened, which is why call this "______________ science," because it focuses on what really happened in the past.

8. Some __________________ events, one in recent history at Mount St. Helens, give us clues as to how water can quickly carve out canyons in a matter of hours.

9. After 150 days of the global Flood, the Bible tell us that water covered the entire earth, and the evidence for this exists all around the world, with 75 percent of the earth's land ____________ comprised of sedimentary layers.

10. We know at the end of the _________ the Ice Age grew quickly, and creation scientists believe it lasted for only a few hundred years.

11. The dam holding back glacial Lake Missoula broke, producing the largest flash flood in history, carving out canyons and scablands in just a few _________, across hundreds of miles.

12. There is a hydraulic process called "______________" where the pressures of high water flow can carve out rock in a matter of minutes.

Chapter 10: Evidence of Dinosaurs

Fill in the blanks with words from the following list:

1,000 6,000 Flood millions sauropod 6 stegosaur varnish

1. The Anasazi left the area around A.D. 1,250, making many drawings on the rock about their life here, including one bearing a startling resemblance to a dinosaur, specifically a ___________, with a long tail and neck, and small head.

2. Those who believe the earth is about __________ years old, as we infer from the Bible, believe these could be the drawings of dinosaurs before they went extinct in this area a few thousand years ago.

3. The Bible says that all land animals were created on day _______ of creation, which would have included the dinosaurs.

4. The earth was much different in climate after the ___________, so the dinosaurs probably didn't come back in the same numbers as before.

5. At an ancient temple in Angkor, Cambodia, a bas-relief carving of a _____________ has been found, which is just over a thousand years old.

6. It's probably only been in the last ________ years that dinosaurs were exterminated into extinction through hunting and changes in the climate.

7. Another compelling claim about dinosaurs living recently is the discovery of soft dinosaur tissue and blood cells, which would have rotted away or fossilized a long time ago if dinosaurs had all died out _____________ of years ago.

8. Several dating methods dismiss the possibility that clever hoaxers created the rock art at Katchina Bridge, including the accumulation of desert _________, weathering, and growth of lichens.

Chapters 11–12: Geology shows judgement and conclusion
Gospel message and closing

Fill in the blanks with words from the following list:

geologists last repent short timeline stegosaur varnish

1. Even though geologic features around the world verify there was a global Flood, secular ________________ continue to deny that truth.

2. The Apostle Peter once said that in the ________ days people will scoff at God's Word, making fun of creation and the global Flood (2 Peter 3:3–6).

3. God's desire is not to destroy His creation, so He waits patiently for man to __________, though His patience will not last forever.

4. A global Flood provides the right conditions to build thousands of arches in a relatively _________ time.

5. The biblical record can be trusted as a book about earth's history with a __________ of only 6,000 years since creation.

Mesa Verde and Chaco Ruins Worksheets

for Use with

Explore Mesa Verde and Chaco Ruins with Noah Justice (Awesome Science Series)

Chapter 1: Opening
Mesa Verde & Chaco Ruins

Fill in the blanks with words from the following list:

Note: Some questions may have more than one answer. Not all of the answers may be used. Some answers may be used more than once.

Billion	Babel	Created	Five	Global	Hundred	Language
Million	Noah	One	Race	Thousand		

1. According to genealogies in the Bible, man was created by God about six ______________ years ago, then a ______________ Flood wiped out everyone, except for ______________ and his family.

2. The Bible states we were all ______________ in God's image and belong to one ______________.

3. Originally there was ______________ language but we also learn that distinct languages came by God at the Tower of ______________ through a supernatural act of God's judgment.

Chapter 2: Mesa Verde Introduction

Fill in the blanks with words from the following list:

Note: Some questions may have more than one answer. Not all of the answers may be used. Some answers may be used more than once.

Abraham Lincoln	Arizona	Billion	Canyons	Colorado	Green
John F. Kennedy	Hundred	Largest	Mountain	Million	New Mexico
Theodore Roosevelt	Palace	Plateaus	Preserve	Smallest	Thousand
Utah					

1. The Southwest part of the United States is an incredible landscape of ______________, ______________ chains, and high ______________.

2. Mesa Verde is located in southwestern ______________.

3. Mesa Verde is Spanish for "______________ table".

4. The park is best known for structures such as Cliff ______________, thought to be the ______________ cliff dwelling in North America.

5. These habitations are thought to have been occupied by ancient natives called the Anasazi about a ______________ years ago.

6. Then in 1906 the national park was established by President ________________________.

7. Mesa Verde was originally set up as an archaeological ______________ so hiking within the park is restricted to the six marked and paved trails.

Discussion Questions

1. Name at least three animals living in Mesa Verde.

2. What insights does Mesa Verde offer?

Chapter 2: Mesa Verde Introduction
What We Find at Mesa Verde

Fill in the blanks with words from the following list:

2	10	100	150	300	600	Largest
Rocks	Smallest	Springs				

Note: Some questions may have more than one answer. Not all of the answers may be used. Some answers may be used more than once.

1. There are about ______________ cliff dwellings preserved at Mesa Verde National Park. Structures range, in size, from one-room storage units to villages of more than ______________ rooms.

2. They are built near ______________, so the people living there had access to water.

3. The Spruce Tree House is the third ______________ cliff dwelling in the park.

4. The Step House is unique because there may be evidence of ______________ separate occupations in the same site.

Discussion Questions

1. Many visitors look at the size of the doorways in Cliff Palace and other cliff dwellings and wonder about the size of the people who once lived here. What does the size of the doorways say about the size of the people?

2. What was one thing that contributed to a short average life span when compared to today?

3. Name three primary construction materials used for the cliff dwellings.

4. Name the three types of material use for the mortar between the blocks.

Chapter 3: Who Built Them?

Fill in the blanks with words from the following list:

700	850	1,000	Ancient	Asia	Canada	Dropped
Drought	Lived	Peaked	Enemy	Moses	Genghis Kahn	
Abraham Lincoln						

Note: Some questions may have more than one answer. Not all of the answers may be used. Some answers may be used more than once.

1. Archeological research has shown they were made by Ancestral Puebloans, also called the Anasazi. Anasazi comes from the Navajo language and means "the ______________ ones" or "______________ ancestors."

2. Some archeologists have proposed that these ancient people migrated from ______________ during the time of ______________.

3. They inhabited Mesa Verde for more than ______________ years. It is thought they primarily lived, on the mesa tops for a good portion of that time, before building the cliff dwellings. It was not until the final 75 to 100 years that they constructed and ______________ in the cliff dwellings for which Mesa Verde is known.

4. There appears to be sporadic, and sometimes long periods of ______________.

5. Research shows the population ______________ in ______________ AD.

Discussion Questions

1. What crops did the inhabitants of Mesa Verde grow?

2. Describe their homes.

3. How do we know in about 930 A.D. people moved back, but now with connections to a vast trade network and commerce?

Chapter 4: Chaco Ruins

Fill in the blanks with words from the following list:

Aligned	Anasazi	Arizona	Aztec	Bonito	Chaco	Drought
Economic	Great	Kivas	Lines	New Mexico	Oklahoma	Pretty
Peru	Primitive	Pueblo	Roads	Salmon	Social	

Note: Some questions may have more than one answer. Not all of the answers may be used. Some answers may be used more than once.

1. Known as the ______________ Canyon complex, the site was the main ______________ and ceremonial center of the ______________ culture.

2. Pueblo ______________, which means " ______________ village" in Spanish, had more than six hundred rooms, numerous two- and three-story buildings, and several ceremonial structures called ______________.

3. Some suggest that a ______________ may have been the reason they abandoned the site.

4. Many of the buildings here are ______________ with the sun, moon, and stars, meaning that these people who built them were not ______________, but highly advanced.

5. The Anasazi constructed massive stone buildings, called ______________ houses.

6. It is estimated that by 1050, Chaco had become the political, ______________, and ceremonial center for the Chacoan culture.

7. ______________ connected the canyon to over a 150 other great houses, including the ______________ Ruins and ______________ Ruins, to the north.

8. Another interesting feature here, in the Chaco Complex, are strange, mysterious ______________ extending ten to twenty miles, into the desert reminiscent to the famous Nasca lines in ______________ .

9. Today, the ______________ Indians of ______________ claim to be the descendants of the ancient Anasazi.

Discussion Questions

1. What did Chaco become the trade center for?

2. Conventional archaeological theories explain the mysterious lines as roads leading to outlying settlements. Why does this seems highly unlikely?

3. Explain the religious beliefs of the Chaco culture.

Chapter 5: Aztec Ruins

Fill in the blanks with words from the following list:

Aztec	Chaco	Commerce	Decades	Homes	Mesa	New Mexico
Monumental	Pottery	Years	Verde	Weeks		

Note: Some questions may have more than one answer. Not all of the answers may be used. Some answers may be used more than once.

1. Just south of the Colorado border is ____________________ and the small town of ____________________.

2. The Aztec Ruins sit between ____________________ Verde and the ____________________ Ruins, so at the height of this civilization there was probably a lot of ____________________ and travel between all three of the regions.

3. In less than three ____________________ work gangs built a ____________________ great house, three stories high and longer than a football field.

4. It appears that an influx of residents of Mesa ____________________ arrived and built additional rooms, remodeled older ones, and used different a type of ____________________.

5. What does the word Pueblo mean? ____________________

Discussion Question

Why do archeologists believe the Anasazi abandoned Aztec, just as they abandoned other villages throughout the Southwest?

Chapter 6: Man's Origins
Man's Origins by Evolution

Fill in the blanks with words from the following list:

Adam	Africa	America	Apes	Asia	Beginning	Billion
Birds	Civilization	Dust	End	Eternal	Garden	God
Humans	Hundred	Middle	Million	Moses	Nimrod	Noah
Nothing	Stars	Thousand				

Note: Some questions may have more than one answer. Not all of the answers may be used. Some answers may be used more than once.

1. According to evolution, the Big Bang happened about 14 ____________________ years ago, causing the universe, to expand and come into existence from ____________________ .

2. Modern ____________________ are believed to have originated around 200,000 years ago, in ____________________ by this secular story.

3. The earth's first major ____________________ at Sumer, in the ____________________ East, was supposedly between 4,000 and 3,000 BC.

Fill in the blanks with words from the following list:

Adam	Bible	Death	Eternal	Garden	God	Life
Moses	Multiply	Nimrod	Noah	Subdue		

Note: Some questions may have more than one answer. Not all of the answers may be used. Some answers may be used more than once.

1. According to the ____________________, about six thousand years ago ____________________ created man in His image on the sixth day.

2. Adam and Eve were given a beautiful ____________________ to live in and told to ____________________ and ____________________ the earth.

3. God originally created man as an ____________________ being, living forever, like God. But God promised that if man disobeyed, ____________________ would be the punishment.

4. Who began taking over cities such as Accad, Erech, Calneh, and Babel?

Discussion Questions

1. Explain why God sentenced man to die for their sin and then cast Adam and Eve from the garden, where they received curses of hard labor, both in work and in childbirth.

2. Explain why people lived shorter lives after the Flood and the Tower of Babel.

3. When we see a rainbow, what does it remind us of?

4. From what three men was the earth populated from after the Flood?

5. Why did God confuse the languages at the Tower of Babel?

Chapter 7: Tower of Babel
Man's Origins by Evolution

Fill in the blanks with words from the following list:

Babel	Destroying	Egypt	Grace	Five	Multiply	Obeyed
One	Rewarding	Rebelled	Spread	Three		

Note: Some questions may have more than one answer. Not all of the answers may be used. Some answers may be used more than once.

1. It was important to God that man ______________ and ______________ out.

2. How many languages were there right after the Flood?

3. Ultimately, God used ______________ to accomplish His purposes. Even though man ______________ against God's command, God showed ______________ and mercy by not ______________ man.

Discussion Questions

1. Within a hundred years after the Flood, where did Noah's descendants settle?

2. Why did the people in Babel build a high tower?

3. Give the approximate year of each event according to the Biblical timescale:

   The Flood: ______________________

   Time of Babel: ______________________

   Babylon: ______________________

   Egypt began: ______________________

   Abraham was called by God: ______________________

4. Who is Egypt named for?

5. What happened after the event at Babel?

6. Where do we get all the amazing people groups and languages we see today?

Chapter 8: Questions of Race

Fill in the blanks with words from the following list:

Adam	All	Beautiful	Biologically	Blood	Characteristics	Darwin
Genetic	Encourage	Evolved	Exterminate	Favored	Genders	Genes
Jesus	No	New	Old	One	Race(s)	Skin
Some	Three	Trivial	Two	Variation		

Note: Some questions may have more than one answer. Not all of the answers may be used. Some answers may be used more than once.

1. ____________________ encouraged the thought that some races were more ____________________ than others and that different races ____________________ at different times and rates.

2. After ____________________ of research, scientists tell us that ____________________ speaking, there is only ____________________ race of humans.

3. Different ____________________ have been used by totalitarian regimes and companies to ____________________ large groups of people, such as Hitler's Germany toward the Jews, Poles, Africans, Slavs as well as Planned Parenthood's early racist attack on dark-skinned African Americans.

4. The Bible describes human beings as "one ____________________", basically saying we are all descendants of the first man, ____________________.

5. Human biological ____________________ is only about 0.012 percent, making "racial" differences absolutely ____________________.

6. ____________________ new genetic information is generated at conception, but a ____________________ combination of already-existing ____________________ information is formed.

7. There are dominant and recessive ____________________. From these genes we get all of our ____________________.

8. Real science in the present fits with the biblical view that ____________________ people are rather closely related—there is only one ____________________ biologically, all coming from ____________________ man and one woman.

Discussion Questions

1. Why were the Australian Aborigines hunted down and killed?

2. What has human genome mapping proven?

3. What creates different shades of skin color?

4. How did different cultures form with certain features becoming predominant within each group?

Establishing the Nations

Fill in the blanks with words from the following list:

Abraham	Adam	Chinese	Common	Earth	Japanese	Japheth
Man	Mexican	Moses	Moon	Noah	Names	Original
Places	Revisions	Shem	Unusual			

Note: Some questions may have more than one answer. Not all of the answers may be used. Some answers may be used more than once.

1. Genesis 5 gives us genealogies from Adam to Noah. It is from these genealogies we can estimate the age of the ____________________.

2. ____________________ was created on day six of creation.

3. Genesis 11 gives us a genealogy, from Noah's son ____________________ to ____________________.

4. Most modern scholars place the date of ____________________ around 2,000 BC, making ____________________ and the Flood around 2,350 BC. This also puts ____________________ at about 4,000 B.C.

5. Some dispute the Table of Nations because there are a few names that seem to be repeated from Genesis 2. But this can easily be solved by understanding that ____________________ and ____________________ often get repeated in honor of those places or people in the past.

6. ____________________ records also describe Nuah with three sons, Lo Han, Lo Shen, and Jahphu.

7. ____________________ is often traced as the ancestor of the European and North Asian nations.

8. Because languages were created, by God, at Babel, the ____________________ names would have had some change between languages and names passed down through generations, might have undergone some slight ____________________, but the root name is still the same.

Discussion Questions

1. What is a genealogy?

2. Why are the genealogy chapters of Genesis some of the more controversial chapters in the Bible, even among Christians?

3. What does 2 Timothy 3:16 tell us?

4. Whose sons and their offspring can be traced to the names of countries, and nations?

5. What did author Bill Cooper find were an incredible match of the Table of Nations when he did an extensive amount of research?

Chapter 9: Origins of Languages
Division of Languages

Fill in the blanks with words from the following list:

100	1,000	4,200	6,900	Adding	American	Animals
Dividing	Fish	Multiplying	Religious	Rocks	Secular	

Note: Some questions may have more than one answer. Not all of the answers may be used. Some answers may be used more than once.

1. There are over ____________________ spoken languages in the world today. Yet, the number of languages emerging from Babel would have been much less than this—as few as ____________________ different original language families.

2. Since Babel about ____________________ years ago, the original language families have grown and changed into the abundant number of languages we have today.

3. Language is one of the main characteristics separating us from ____________________.

4. ____________________ linguists can't explain how so many languages formed.

5. God's plan at Babel was to spread out man, over the earth, by ____________________ the people by language to thwart their sinful rebellion.

Discussion Questions

1. What is linguistics?

2. What does 2 Timothy 3:16 tell us?

Chapter 10: Where did they go? Where Did Anasazi Come From?

Fill in the blanks with words from the following list:

Abraham's	Assumptions	European	Families	Geographical	Ham's	Japheth's
Jewish	Languages	Minor	Mexican	Native	Noah's	Shem's
Trusted						

Note: Some questions may have more than one answer. Not all of the answers may be used. Some answers may be used more than once.

1. As a general trend, ____________________ descendants went toward Europe and North Asia, ____________________ went toward Africa and the Middle East, and ____________________ mainly kept in the Middle East and Far East, although there are some exceptions to this, in each line.

2. When it comes to North America, and even portions of South America, ____________________ settlers have usurped many names and places and provided new names. Most of the ____________________ names have been lost and forgotten.

3. ____________________ names are a big key to determining names of people and nations from the past. Yet, they may be derived from various descendants well beyond the names given in Genesis 10.

4. Due to wars and migrations of people, ____________________ often intermingled with other families.

5. With regards to Native North Americans, a handful of language ____________________ seem to dominate.

6. The people we know as ____________________ Americans were not the first people on this continent.

7. The discovery at a cave, in Northwest Oregon, yielded human coprolite, human feces. It was found to contain mitochondrial DNA ____________________ ethnic groups from Siberia and East Asia.

Discussion Questions

1. What do language families consist of?

2. Natives of America passed their language down orally. What did this allow for?

3. What evidence about the Clovis people has surfaced?

4. What does the post-Babel model of all language/people groups say about migration to the Americas?

5. What parts of the world took longer to populate?

Where Did Anasazi Go?

Fill in the blanks with words from the following list:

Australia East Japan Mexico Modern North South

West

Note: Some questions may have more than one answer. Not all of the answers may be used. Some answers may be used more than once.

1. The Anasazi went ____________________.

2. The Shoshoni language, in the Utah-Nevada region, was understood by all the tribes from ____________________, without difficulty.

3. It has also been found that the Anasazi are related to ____________________ tribes including the Paiute, Hopi, Pima, Yaqui/Apache, Tepehuan, Kiowas, and Mayos.

Discussion Question

What did Catholic missionaries in the 1850s establish?

Chapter 11: God's Judgement
Christ, the Door

Fill in the blanks with words from the following list:

Abraham	Adam	Ark	Eve	Fire	Flood	Forever
Earth	Forever	Languages	Mary	Punishment	Rebellion	Repent
Save	Sin	Wives				

Note: Some questions may have more than one answer. Not all of the answers may be used. Some answers may be used more than once.

1. According to Genesis 3, __________________ and __________________ were the first to sin.

2. In Noah's day, man's rebellion against their Creator, had gotten so bad, that, God destroyed the entire earth, in a __________________ .

3. Even then, man still wouldn't obey God to spread out across the earth, so He confused their __________________ .

4. Over and over, we see in the Bible that God is patient with man's __________________, but He won't let sin continue __________________ .

5. Even today, people all around the __________________ are rebelling against the Creator. They would rather deny there is a God and follow in their own __________________.

6. When we were powerless to do anything to __________________ ourselves, God loved us so much that He sent his Son, Jesus Christ, to die in our place, to take the __________________ for our rebellion and sin.

7. Even with this great gift we've been given, many around the world have rejected God's grace because they love their __________________ too much.

8. God offers every man the chance, to __________________ of his sins, and turn to Him. We are encouraged, in the Bible, to not put off this decision, but come back to God, today.

9. Just like the door of the __________________ was open, for a time, for anyone who wanted to be saved from the Flood, the door is now open through Jesus Christ to repent of our rebellion and sin. And just as there was a time when the door, of the ark, was sealed, the opportunity to come to God won't last __________________.

Discussion Questions

1. What does the Biblical account of man on the earth match?

2. What does the Table of Nations match?

3. What does the account of the Tower of Babel match?

4. What does the evidence of the movement of man, from the starting point of Mesopotamia, across the earth match?

5. What does the Bible say our rebellion against God deserves?

6. What does Romans 6:23 say?

If you have not humbled yourself, repented of your sins, and turned your heart to Him, then we invite you to do that today.

Chapter 12: Conclusion
Closing Scene

Fill in the blanks with words from the following list:

Bible	Confusion	Disaster	Dispersion	Dying	Evolutionists	Exodus
Genesis	Jeremiah	Mysteriously	Relent	Scientists	Thriving	Turns

Note: Some questions may have more than one answer. Not all of the answers may be used. Some answers may be used more than once.

1. Mesa Verde, the Chaco Ruins, and Aztec ruins remind us that a once ____________________ culture used to live, in the southwest, but they ____________________ disappeared and meshed into other cultures, particularly to the south.

2. The ____________________ of man from Babel, by the ____________________ of their languages, gives us an idea how God caused man to move out, from the Middle East, into the rest of the world.

3. The ____________________ gives us clues about why cultures come and go.

4. In the book of ____________________ God says, "The instant I speak concerning a nation and concerning a kingdom, to pluck up, to pull down, and to destroy it, 8 if that nation against whom I have spoken ____________________ from its evil, I will ____________________ of the ____________________ that I thought to bring upon it."

Discussion Questions

1. What will God do if a nation honors Him?

2. What will God do if a nation rebels against Him?

3. What does archaeology indicate about ancient cultures in the Southwest? What was the result of this?

4. What lessons should we learn from the ruins?

Rocky Mountain National Park Worksheets

for Use with

Explore Rocky Mountain National Park with Noah Justice
(Awesome Science Series)

Chapter 2: History and Features
Rocky Mountain National Park

Fill in the blanks with words from the following list:

British Columbia	New Mexico	North America	Hundred
Thousand	Billion		

Note: Some questions may have more than one answer. Not all of the answers may be used. Some answers may be used more than once.

1. Where are the Rocky Mountains located?

2. The Rocky Mountains stretch over a _____________ miles, from ____________________________ to ______________________.

Discussion Questions

1. How are sedimentary rocks formed?

2. What is Catastrophic Plate Tectonics?

Chapter 2: History and Features
Rocky Mountain National Park History and Features

Fill in the blanks with words from the following list:

72 415 1915 4,000 12,000 Abraham Lincoln

Woodrow Wilson

Note: Some questions may have more than one answer. Not all of the answers may be used. Some answers may be used more than once.

1. When was the Rocky Mountain National Park established?

2. Who established the Rocky Mountain National Park?

3. How many square miles does the Rocky Mountain National Park cover?

4. The elevation range of the park is _______________feet to over _______________ feet.

5. How many peaks does the park have?

Explore Rocky Mountain National Park	Rocky Mountain National Park DVD	Day 164	Worksheet 2	Name

Chapter 3: Mountain Building

Fill in the blanks with words from the following list:

Aleutians Alps Appalachians Cascades Himalayas Grand Tetons in Wyoming

Hawaiian Islands Rockies Black Hills of South Dakota

Note: Some questions may have more than one answer. Not all of the answers may be used. Some answers may be used more than once.

1. Give four examples of folded mountains. ______________ ______________ ______________ ______________

2. Give an example of domed mountains. ______________________________

3. Give an example of fault block mountains. ______________________________

4. Name three volcano linear chains located around the Pacific Rim. ______________ ______________ ______________

Discussion Questions

1. How many continents are mountains found on?

2. How are folded mountains formed?

3. How do domed mountains form?

4. How are fault block mountains formed?

5. How are volcanoes formed?

6. What formed most of the mountains and in what time span?

	Explore Rocky Mountain National Park	Rocky Mountain National Park DVD	Day 165	Worksheet 3	Name

Chapter 4: Secular Ideas

Fill in the blanks with words from the following list:

Four centimeters		Four feet	Four inches	Four millimeters	Fossils
Granites	Metamorphic rocks	Marine fossils	Mt. Everest	North American	Pacific
Rockies	Sedimentary rock	South American		Water	

Note: Some questions may have more than one answer. Not all of the answers may be used. Some answers may be used more than once.

1. What are the Rockies full of?

2. What are below the sedimentary rocks in the Rocky Mountains?

3. What is the tallest mountain in the world?

4. How much does Mount Everest grow in height a year?

5. Name three plates.

6. What plate began pushing down below the North American Plate and what mountain range formed as a result?

7. What is the top layer of Everest?

8. What was the top layer of Everest formed by?

Discussion Questions

1. What are metamorphic rocks?

2. What is subduction?

Chapter 5: Biblical Model
Biblical View on Rockies Formation

Discussion Questions

1. Do Creation scientists and secular scientists agree about the mechanism for creating the Rockies?

2. What do Creation scientists and secular scientists disagree about?

3. What is rapid horizontal divergence of tectonic plates?

4. What did geologists believe before Pellegrini developed his idea of is rapid horizontal divergence of tectonic plates?

5. What did creationist geologists in the 1990's call Pellegrini's idea for quick movement of the earth's plates during the Flood?

6. Explain how a pre-Flood earth was made up.

7. How does Catastrophic Plate Tectonics propose the Flood started?

8. According to Catastrophic Plate Tectonics, how fast did the plates move, and what did this lead to?

9. What did the mantle convection affect, and what did it induce?

10. What are flood basalts?

11. What caused intense global rain according to Catastrophic Plate Tectonics?

12. What caused the sea level to rise so that the ocean water swept up onto and across the continents?

13. What resulted in sedimentary rock layers covering regions and entire continents?

Chapter 6: Catastrophic Plate Tectonics Examined

Fill in the blanks with words from the following list:

Continental edges Cooler Grand Canyon Granites Ice Age Large earthquakes

Metamorphic rocks Northern Arizona Subducted Warmer Zion Canyon in Utah

Note: Some questions may have more than one answer. Not all of the answers may be used. Some answers may be used more than once.

1. What caused tidal waves?

2. Where can we see erosion of these sediment layers?

3. When the pre-Flood, cold denser ocean floor ____________________ under the moving continents and plunged into the mantle, it also tried to drag the ____________________ ____________________ with it.

4. Give two examples of the rocks that got pushed up because the subduction pre-Flood floor slab sank below creation week rocks foundational to the North American Plate.

5. The huge geologic activity, in the oceans, during the Flood, would have heated up the water worldwide. This resulted in a significantly ____________________ climate, just after the Flood.

6. Large storms dumped huge amounts of rain and snow on the continents, causing heavy erosion, and the quick development of the ____________ ____________ toward the poles.

Discussion Questions

1. What slowed and eventually stopped the catastrophic plate motion?

2. What caused the Rockies to rise higher?

3. What was happening to the ocean basins as the mountains were rising on the continents and the flood waters were quickly receding?

4. What carved huge canyons?

Chapter 7: Rate of Erosion

Fill in the blanks with words from the following list:

23	42	2,300	5,000	9,000	Bible	Billion
Charles Darwin		Erosion	Hundred	Ice Age	James Hutton	
Larger	Million	Mountains	Planation	Surfaces	Precipitation	Regional
Smaller	Thousand					

Note: Some questions may have more than one answer. Not all of the answers may be used. Some answers may be used more than once.

1. Who suggested the earth was immensely old in 1785?
2. Before that, many scientists believed the earth was young, using the ______________ as their guide.
3. One process that indicates a younger earth is the process of ______________.
4. If the earth were even hundreds of millions of years old, ____________________ should not exist.
5. The Bible states the highest peak from the earth before the Flood was covered by over _______ feet of water.
6. If every continent was laid flat and the ocean basins filled in so that everything was level, there is enough water to cover the entire earth up to 1.7 miles. That's about _________ feet.
7. What do we see at the tops of mountains all around the world that can only result from quick, catastrophic processes?
8. After the Flood, the oceans were very warm due to volcanic activity and ejection of subterranean water, mentioned in the Bible as "Fountains of the Great Deep." These warm oceans would have caused massive ______________________________ across the continents, making the rate of erosion much greater than it is today.
9. This massive ______________________________ would have also brought on an ______________________, in a matter of a few ____________________ years.
10. At the end of the Ice Age the melting ice caused large ____________________ floods, such as the Missoula Flood in the Pacific Northwest.

11. The rate of erosion was much ______________________ in the past.

12. When you include the level of erosion during and after the global Flood, you can easily show billions of years of erosion in just a few ________________________ years.

13. The ____________ can be trusted as earth's history book.

Discussion Questions

1. Secular scientists say the process of erosion and mountain building is ongoing so that as mountains are eroded, they continue to uplift and build, so the effect of erosion is not as noticeable. What is the major problem with this explanation?

2. Scientists have figured human activity can increase the erosion rate 2 to 2.5 times. Does this explain the problem of erosion over long ages? Why or why not?

3. What does Psalm 104:8 tell us?

4. How is a Planation Surface formed?

Chapter 9: The Great Sand Dune

Fill in the blanks with words from the following list:

50	350	750	Billion	Climate	Common	Crowded
Desolate	Giraffes	Hippos	Hundred	Ice Age	Man	Million
Mountain	Sand	Silt	Thousand	Unusual		

Note: Some questions may have more than one answer. Not all of the answers may be used. Some answers may be used more than once.

1. The dunes at Great Sand Dunes National Park in Colorado rise about __________ feet from the floor of the San Luis Valley and cover about 19,000 acres.

2. The Great Sand Dunes is at 8,200 feet above sea level. It is ____________________ to find sand dunes this high in the mountains.

3. Coastal dunes tend to have more minerals than ____________________ dunes.

4. We know that sand dunes accumulate during times of ____________________ change.

5. Not long after the Flood, the ____________________ developed.

6. At the end of the Ice Age, there was a drier period where wind-blown ____________________ was carried across continents.

7. The Sahara Desert used to be a lush area with ____________________, ____________________ is and habitations of ____________________ but now is covered in ____________________ and ____________________.

8. Given the right conditions after the Flood, vast areas of dunes could have developed quickly, in just a few ____________________ years.

Chapter 11: Devils Tower

Fill in the blanks with words from the following list:

4	7	11	312	1896	1906	1976
1,000	3,000	5,000	Abraham Lincoln		Eroded	Height
Igneous	John F. Kennedy		Magma	Power	Sediment	Sedimentary
Theodore Roosevelt		Utah	Water	Wyoming		

Note: Some questions may have more than one answer. Not all of the answers may be used. Some answers may be used more than once.

1. As the Rockies were rising quickly at the end stages of the Flood, huge sheets of ____________________ were running across the continents and thousands of feet of top ____________________ layers were being ____________________.

2. The Devil's Tower is found in Northeastern ____________________.

3. The Devil's Tower truly is a testament to the receding flood water's ____________________.

4. Devil's Tower was declared a national monument in ____________________, by President ____________________.

5. Devil's Tower is made up of spectacular vertical columns of ____________________rock with five or six sides each. At the base, these columns are about ____________________ feet wide and decrease to ____________________ feet near the top.

6. Scientists have been able to determine the surrounding land was nearly ____________________ feet higher than it is today.

7. In the case of Devil's Tower, the____________________ never reached the surface.

Discussion Questions

1. Give two theories on how the Devil's Tower was formed.

2. What do the Devil's Tower columns, composed of rock and containing fine-grained minerals, suggest about how they were formed?

3. How was the Devil's Tower exposed?

4. What does the current erosion rate say about secular theories of the Devil's Tower?

Chapter 12: Conclusion

Fill in the blanks with words from the following list:

Abraham	Assumptions	Billions	Flood	Ice Age	Judgment	Major
Millions	Minor	Moses	Noah	Quick	Quickly	Slow
Slowly	Thousands	Trusted				

Note: Some questions may have more than one answer. Not all of the answers may be used. Some answers may be used more than once.

1. Rocky Mountain National Park is a great place to learn how the ________________ provided the right conditions to form mountains ________________ around the world.

2. Secular methods for dating the rocks can't be ________________ because of the many false ____________ made by scientists.

3. Erosion rates show that the Earth is ________________ years old.

4. Geologic features, such as Devil's Tower and the Great Sand Dunes, show ________________ formation and erosion due to the effects of the Floodwaters, and the climate changes, after the ________________ , also caused by the ________________ .

5. The ________________ had a ________________ impact on the earth we see today. The Bible says the Flood was God's ________________ on sin and rebellion, back in the days of ________________.

With everything happening in our world today, and where man is headed, God promises another judgment on man's rebellion and sin, this time by fire. He loves His creation but cannot let it go on the way it is, toward destruction. But He promises if we repent of our evil and turn to His Son, Jesus Christ, and trust in His death on the cross as THE payment for our sin, then God will forgive us, and we will walk with Him, forever, in a new heavens and new earth.

If you have not done so already, we encourage you to repent of your sin and turn to Jesus Christ today.

Test Section

True North Grand Canyon / Concepts & Comprehension	Test 1	Scope: Grand Canyon	Total score: ____of 100	Name

Define: (5 Points Each Answer)

Schist: ______________________________

Granite: ______________________________

Unconformity: ______________________________

Land bridge: ______________________________

Tapeats Sandstone: ______________________________

Biological mutualism: ______________________________

sheet erosion: ______________________________

sapping: ______________________________

riparian: ______________________________

ecosystem: ______________________________

Short Answer Questions: (5 Point Each Answer)

1. Both the Havasupai and Hualapai, tribes native to the Grand Canyon, have retained legends of a ____________ ____________.

2. What kind of rock makes up the red cliffs seen throughout the canyon?

3. Archaeologists have determined the Tusayan Village was inhabited around 1,000 years ago by what people?

4. Why did the Anasazi disappear from the Grand Canyon?

5. What is the highest point in Arizona?

6. When was the Grand Canyon declared a national monument?

7. What is the difference between a contact and an unconformity?

8. How would a rapid flow of water during a catastrophic process explain the lack of rockfall at the bottom of the Canyon?

9. Billions of fossilized ________________ are entombed in a seven-foot-thick layer near the base of the massive Redwall Limestone formation.

10. What bird has been specifically designed to swim on the water and in the water and to walk along the bottom of the Colorado River?

First Semester Test / Concepts & Comprehension	Test 2	Scope: First Semester	Total score: ____of 100	Name

Discussion Questions: (5 Point Each Answer)

1. Why is it important to understand how water flow and erosion can create land features in trying to see if the Colorado River could have carved the Grand Canyon?

2. What is the process known as erosion?

3. Why do these scientists not agree on one explanation for how the Grand Canyon formed if everyone looks at the same evidence?

4. Why are secular scientists still debating the merits and problems of different theories in an attempt to solve the riddle of the origin of the Grand Canyon?

5. Discuss why biocrusts are often called the "glue of the desert."

6. What is volcanic tuff?

7. What effect does fire have on Yellowstone in terms of plant life?

8. In which part of Yellowstone Park are bison normally found?

9. What is the Painted Desert?

10. How do volcanoes and floods impact the landscape?

11. What is a meteor?

12. What process do secular scientists think occurred to make the deep valleys?

13. How does the biblical story of Noah and the Great Flood offer a better explanation for these geologic features?

14. How do the biblical genealogies in Genesis help us calculate the age of the earth?

15. Having seen this type of blast for the first time, scientists were now able to find 300 other volcanoes on earth that they were able to explain based on the observations from Mount St. Helens. What other volcanoes are known to have erupted the same way Mount St. Helens did?

16. What descriptions in Psalm 104:23 that are hinting at volcanoes speaks to the power of God?

17. A proper understanding of the evidence after actual observations of the rapid canyon formation at Mount St. Helens led many researchers to conclude that to carve canyons of large magnitude you need a lot of water in a short period of time, not the small rivers over millions of years. Why was it so important that scientists be able to see these events actually take place in understanding the evidence left behind at other locations?

18. There are a large variety of fossils at the John Day Fossil Beds. The Bible tells us that God sent the Flood to destroy all land-dwelling, air-breathing life, except for Noah, his family, and the animals on the ark. Do the fossils found here support that?

19. What do the presence of flying creatures in the fossil record at John Day reveal about the scope of this catastrophic event?

20. How does a global flood explain the presence of fossils from environments different than that in which they were found?

True North Yellowstone / Concepts & Comprehension	Test 3	Scope: Yellowstone	Total score: ____of 100	Name

Define: (5 Points Each Answer)

Caldera: ______________________________

Plateau: ______________________________

Thermophile: ______________________________

Supervolcano: ______________________________

Erosion: ______________________________

Biodiversity: ______________________________

Fault scarp: ______________________________

Contact: ______________________________

Short Answer Questions: (5 Point Each Answer)

1. At what elevation will you find Yellowstone National Park?

2. Yellowstone has over ________________ geothermal features within the park.

3. What have uniformitarian thinkers discarded in their theory of millions of years of geologic time?

4. Some critics of a global flood suggest there is not enough water on the earth to cover all the mountains. Yet today, if the topography of the earth were flattened to form a perfectly round sphere, the planet would be covered by water over ______________ miles deep.

5. When did the first eruption in Yellowstone likely occur?

6. The different colors in the hot pools and terraces at Mammoth Hot Springs are caused by various heat-loving ________________________.

7. What kind of rocks make up the Teton Mountains?

8. What sculpted the grooves in the bedrock of Cascade Canyon?

9. What do clouds, glaciers, streams, and lakes all have in common?

10. What percentage of our planet is covered with water?

Second Semester Test / Concepts & Comprehension	Test 4	Scope: Second Semester	Total score: ____of 100	Name

Discussion Questions: (5 Point Each Answer)

1. What is the name of the mountain range just south of Yellowstone?

2. What four types of thermal features are found at Yellowstone?

3. What happened during the Missoula Flood in the Pacific Northwest?

4. How do we know large regional floods, like the Missoula Flood, occurred?

5. What are erratic boulders?

6. How does sin affect secular science?

7. All dinosaurs are found in ______________________________ layers of sand, silt, and mud, which would seem to indicate they were buried quickly in a flood.

8. Give two reasons why iridium in a meteorite from an asteroid does not make sense as the sole source for the iridium found in this layer at Yellowstone.

9. Why did Phil Anderson decide to sell the land where the Mammoth Site is?

10. How does carbon 14 enter animals' bodies before they become fossils?

11. Why is it impossible to use radiocarbon dating for things that are supposed to be millions of years old?

12. Name at least three animals living in Mesa Verde.

13. What insights does Mesa Verde offer?

14. Many visitors look at the size of the doorways in Cliff Palace and other cliff dwellings and wonder about the size of the people who once lived here. What does the size of the doorways say about the size of the people?

15. What is Catastrophic Plate Tectonics?

16. How are folded mountains formed?

17. How are volcanoes formed?

18. What carved big canyons?

19. How is a Planation Surface formed?

20. What was happening to the ocean basins as the mountains were rising on the continents and the flood waters were quickly receding?

Map of National Parks and Monuments in the Western United States

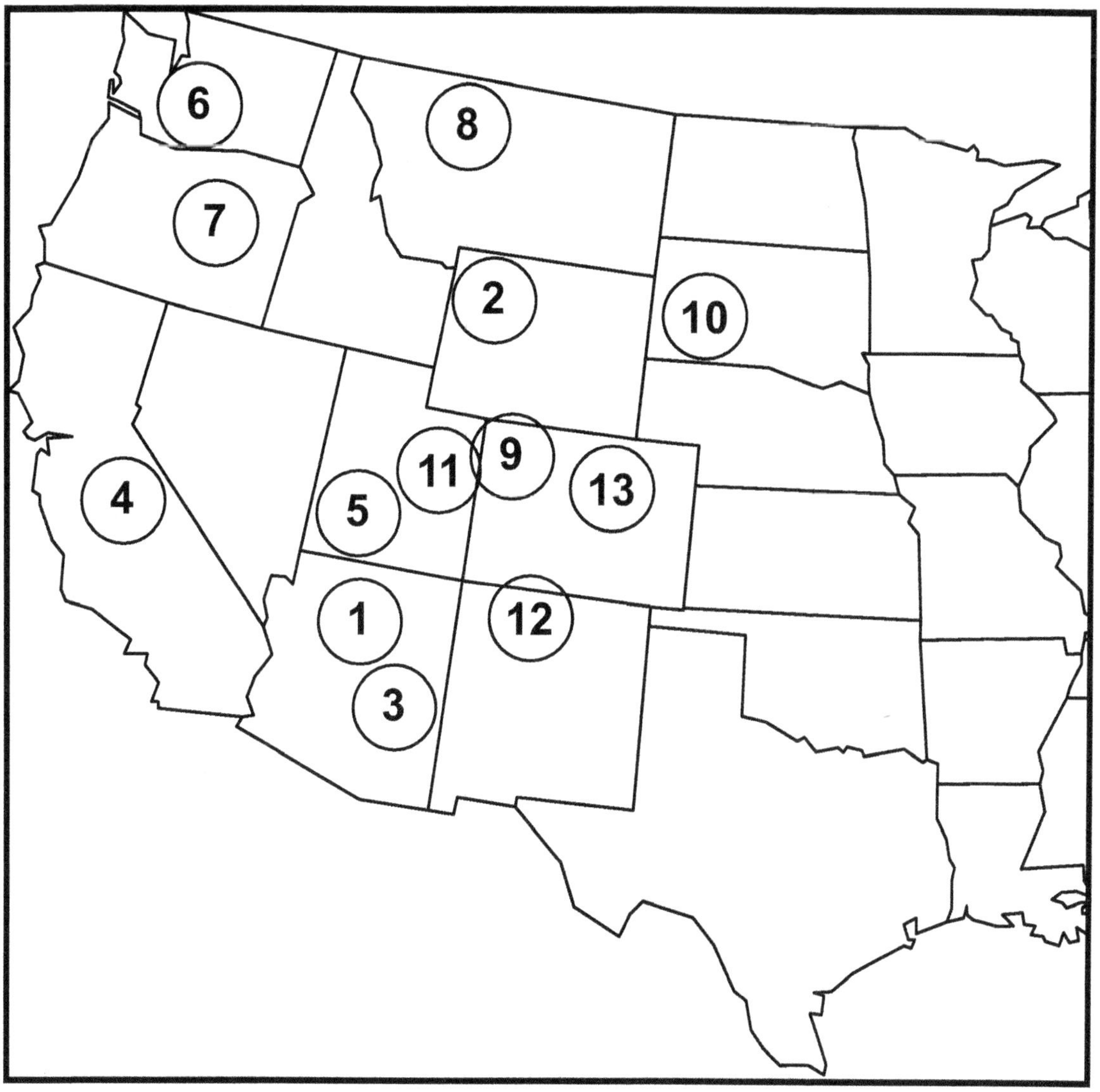

1. ______________________________
2. ______________________________
3. ______________________________
4. ______________________________
5. ______________________________
6. ______________________________
7. ______________________________
8. ______________________________
9. ______________________________
10. ______________________________
11. ______________________________
12. ______________________________
13. ______________________________

Map of National Parks and Monuments in the Western United States

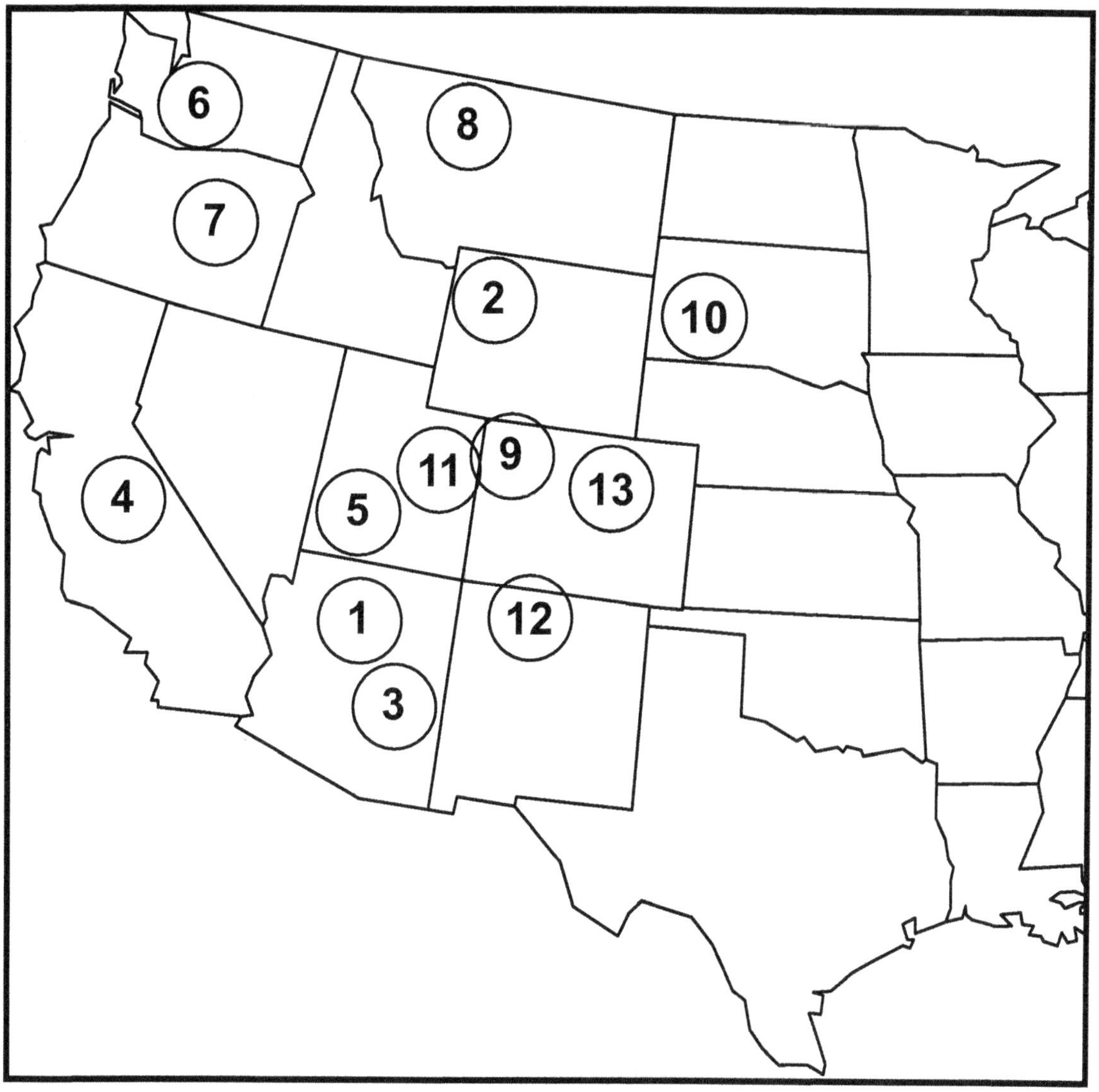

1. __________________________________
2. __________________________________
3. __________________________________
4. __________________________________
5. __________________________________
6. __________________________________
7. __________________________________
8. __________________________________
9. __________________________________
10. __________________________________
11. __________________________________
12. __________________________________
13. __________________________________

Answer Keys

Worksheet 1

About the Grand Canyon

1. Western, scenic
2. Arizona, moon
3. Colorado, Sea
4. Native, expedition
5. Sedimentary
6. Basement, schists
7. Millions
8. Slow, processes, secular humanism, biblical
9. Evolutionary, present, Noah's
10. Canyon, Genesis
11. Worldview, reinterpret, evidence

Discussion Questions

1. Your worldview represents your starting point for when you look at the world around us. With a secular worldview, you are assuming millions or billions of years formed the world we see today. The evidence doesn't always support this secular timeline. This is why some scientists are having to reinterpret their thinking, because they realize a proper interpretation of the evidence just doesn't match up with millions of years.
2. Because the small river is not enough to have carved the canyon, even if you gave it millions of years. That is why catastrophe on the scale of the Flood makes more sense, and fits the evidence we see today.
3. Native Americans first dwelt here, but in 1540 explorers with Coronado were the first known Europeans to see the canyon.

Worksheet 2
Evidence: Colorado River Plausibility Questions

1. False
2. True
3. False
4. False
5. False
6. True
7. True
8. True
9. False
10. True
11. True

Discussion Questions

1. Secular scientists say that the Colorado River carved the Grand Canyon over millions of years — yet the evidence does not support this idea. By tracing where the river begins and ends, and all the land it goes through, you can see that isn't possible. The river doesn't have the volume of water needed to do it even over a long period of time, there is no evidence the Kaibab Plateau rose at the time needed, and a river cannot run uphill — and certainly not the 4,000 feet needed. Most scientists now think that the Kaibab Plateau rose before the Canyon was cut — so something else must have been happening.
2. Gravity is the force that keeps your feet and other things on the ground. Rivers always erode downward because of gravity.
3. Erosion is a process by which water or another force can wear away the surface of something else — like water making a rock smoother.

Bonus Activity:

Answers will vary. This activity can be in the form of an essay, a discussion, or additional research opportunity.

Worksheet 3
The Biblical Record

Discussion questions:

1. Answers will vary. Example: It tells us that sin is a part of us, and that the world can help influence that. If sin existed only in your heart, it would be infecting only you. But sin is everywhere, and it has been around since the beginning of history. Sin can influence what we

say, do, and think, and in our interactions with one another.

2. Sin is what keeps us from having a relationship with God – and God is holy and without sin. When Adam and Eve sinned, they destroyed the relationship they had, and indeed all mankind forever, with God. But God created a plan to save people and be able to have a relationship them. He sent His Son, Jesus, to die as a sacrifice for our sin. If we accept Jesus into our hearts, we can be saved and have a relationship with Him and God.
3. Noah found favor with God because he was a righteous man in a world full of people who were wicked, evil, and violent.

Worksheet 4
Looking at the Evidence – the Canyon Walls

1. Layers, basement
2. Sedimentary
3. Water, sand
4. Sediments, pressure
5. Below, volcanic, metamorphic
6. Megascopic, creation
7. Canyon, oceans
8. Recede
9. Scavengers, decomposition
10. Sea level
11. Crust, recessional

Sequence of the Flood:

7, 4, 6, 13, 3, 1, 9, 10, 11, 8, 2, 12, 14, 5

Worksheet 5
It's the Timing!

1. Bent, wet
2. Sequence
3. Uplifting, sinking
4. Marine
5. Fossils
6. Boundaries, deterioration
7. Burrowing
8. Root systems
9. Time
10. Bottom
11. Pebbles, fast
12. Quickly
13. Crossbeds, geologic
14. Sandstone, desert

Worksheet 6
Dunes

1. True
2. False
3. False
4. True
5. False
6. True
7. True
8. True
9. True
10. False
11. True
12. True
13. False
14. True
15. False
16. True
17. False
18. False
19. True
20. True

Discussion Questions

1. No. They look at the same evidence, but each has a different starting point in looking and studying the evidence.
2. Because of their worldviews. Secular science says the world took billions of years to appear as it does today because they say changes took place in a slow way over millions of years. Creation scientists say the world is very young based on the biblical record and the evidence they see in

the world in the context of the biblical event of the Great Flood.

3. It shows that the history found in the Bible is true – and that the Bible is more than just a book of religious stories; it is a history book in and of itself. In many instances we see or learn something in the Bible, but it may have taken science hundreds of years to confirm it.

Worksheet 7
Receding Floodwaters

1. Kaibab
2. Evidence, weather, post-Flood
3. Hurricanes, stabilizing, polar
4. Rainfall
5. Limestone, leaches, weaken, saturated
6. Grand Canyon
7. Greater, Canyonlands, miles
8. Soft

Worksheet 8
Examples at Mount St. Helens

1. Answers will vary. For example: We know when and how these events happened. We know the results of this event and scientists have been able to study it, rather than relying on theories or inaccurate dating methods. And from what we can see and study, it shows remarkable answers about how other places may have formed.
2. Answers will vary. For example: Dormant volcanoes are ones that are not showing activity, or actual, eruptions – often for a very long period of time. It is very important to be able to see the events as they happened – to know what the area was like before, during, and even after the event. Often scientists with a biblical worldview are told that they do not understand science and its processes, when in reality the creation scientists and secular scientists are looking at the same evidence and interpreting much of it the same way, but having a different starting point gives them a different perspective on what they see. What happened at Mount St. Helens and all it affected, despite the dating of rocks saying it happened a long time ago, helps to prove that secular timelines can be flawed and very inaccurate if they are relying on certain data.
3. Answers will vary. For example: These are processes that were taking place in different places in the world while the Flood was taking place. You can see how one catastrophic event can have a huge impact on the area in a lot of ways.
4. Answers will vary. For example: The dam was not going to be strong enough to probably hold in place during an eruption. You have a situation where you have a lot of water being gathered in the form of ice and snow, and an active volcano. It was just a matter of time before the dam was destroyed. And this process can also help explain some other features we see in the world today.
5. Answers will vary. For example: the landscape was changed due to the creation of canyons and other geologic features in a very short period of time compared to how it looked before. Also, you can see how the event devastated the landscape, but even within a short time, you see the area trying to recover with vegetation and animals eventually returning to the area.
6. Answers will vary. For whichever view you take, be sure to give supporting evidence. For example: a lot of water over a little time. We have seen examples of this kind of event at Mount St. Helens, and there are too many factors that don't fit for a little water over a lot of time having carved the Grand Canyon – such as not enough volume of water, the forces of gravity, and the Kaibab Plateau.

Worksheet 9
Examples at Glenn Canyon Dam and Cavitation

1. Answers will vary. For example: Many think that water can only change certain things like rock or whole landscapes gradually over a long period of time. But this shows that with a lot of water and a catastrophic event that takes a small amount of time, it can be even more destructive than one thinks. It shows that the events of the Great Flood could have created the conditions and geologic features we see in the world today.
2. Answers will vary.
3. Answers will vary. For example: It is a situation

in which water can tear apart almost any surface or material when enough pressure and flow is taking place – and that includes on a larger scale. Studies can reveal the effects of such an event as well as having real-life examples.

4. Answers will vary, but should include an affirmative or yes answer.
5. Answers will vary. Secular sources of information are readily available online and in your local library. If you are looking for creationist perspective, you can find information online at sites like Answers in Genesis or the Institute for Creation Research among other sites if books are not available at your local library. An example of an article can be found at: http://www.answersingenesis.org/articles/wog/grand-canyon

Worksheet 10
Example of Mars

1. False
2. True
3. False
4. True
5. True
6. True

Bonus activity:

Answers will vary.

Discussion question:

Answers will vary. For example: Because they choose to believe a different timeline for events here on Earth than for Mars. Secular science is locked into the idea that the earth is billions of years old. With that as a starting point, they have to follow that idea in terms of all the events that follow. They also think they know much more about what events have taken place on Earth than they do on Mars, where their ability to study is nearly impossible.

Your Guide to the Grand Canyon — Worksheet Answer Keys

Worksheet 1, Foreword – Section Four

Terms to Know and Spell

schist — a type of metamorphic rock that often contains reflective minerals such as mica.

granite — a common and widely occurring type of igneous rock, commonly made up mostly of light-colored minerals such as quartz and feldspar.

unconformity — a buried erosion surface separating two rock masses or layers of different ages, indicating that sediment deposition was not continuous; in the evolutionary model this also represents a lengthy interval of missing time.

Short Answer Questions

1. dead things, water, earth
2. pictures, footsteps
3. horse, donkey
4. endurance
5. carefully, good

Discussion Questions

1. A worldwide Flood is rejected by those who do not believe the Bible is the ultimate source of truth and thus the scientific evidence is being filtered or interpreted through worldviews that do not include the biblical record.

Worksheet 2, Section Five, Part 1

Terms to Know and Spell

land bridge — land connecting continents or landmasses that permits the migration of animals, often subject to temporary or permanent submergence.

Tapeats Sandstone — the lowest horizontal layer in the Grand Canyon, which sits directly on top of the basement rocks.

Great Unconformity — the contact point between the basement rocks of the Grand Canyon supergroup and the sedimentary layer above it.

Short Answer Questions

1. 4,600

2. 277, 18

3. basement rocks

4. 9 ½

5. **bio-crust** — the dark, lumpy crust covering much of the untrampled soil in the Grand Canyon, a mixture of lichens, mosses, microscopic fungi, bacteria, and green algae.

Discussion Questions

1. Biocrusts are often called "the glue of the desert" because they can absorb 10 times their volume in water and are especially important in the storage of rainfall. It is also a key factor in reducing both wind and water erosion.

Worksheet 3, Section Five, Part 2

Terms to Know and Spell

biological mutualism — a biological interaction between individuals of two different species, from which both individuals derive a benefit.

granite gorge — three areas of the Canyon (upper, middle, and lower) where the basement rocks are exposed at the river level.

nautiloid — a category of marine mollusk. They are cephalopods, which includes modern creatures such as the octopus, squid, and cuttlefish. Nautiloids are characterized by a chambered shell that houses the soft parts of the animal.

Short Answer Questions

1. kangaroo rat
2. John Wesley Powell
3. It carries pine seeds long distances in a specially designed pouch under its tongue.
4. Great Flood

Discussion Questions

1. The directional orientation of the nautiloid fossils suggests they were sorted and buried in a huge underwater landslide resulting in the rapid deposition of the nautiloid layer.
2. Various global flood legends lend credence to the history presented in the Bible. It is believed that the story of the Flood was verbally passed from one generation to the next after people dispersed from the Tower of Babel. Some aspects would have been lost or altered. However, each story shares remarkable similarities to the account of Noah in the Bible. For additional information see *Flood Legends* by Charles Martin (Master Books, 2009).

Worksheet 4, Section Five, Part 3

Terms to Know and Spell

crossbeds — an inclined arrangement of thin sedimentary layers in a larger horizontal layer that indicates the direction of the wind or water currents in the depositional environment.

Tower of Babel — (as used in the Bible) a structure built by man attempting to reach the heavens that caused God to confuse the previously uniform language of man, ultimately leading to the dispersion of people across the globe.

butte — a conspicuous hill with steep sides and a flat top. The top is usually a cap-rock of resistant material. This structure is frequently an erosional remnant in an area of flat-lying sedimentary rocks.

mesa — an area of high land with a flat top and two or more steep, clifflike sides. Mesas are larger than buttes and smaller than plateaus.

gorge — a deep, narrow passage with steep rocky sides; a ravine.

Short Answer Questions

1. seeds
2. the Monarch butterfly
3. redwall limestone
4. desert lupines
5. the Anasazi

Discussion Questions

1. They were disobeying God's command to spread out over the earth. Additionally they thought if only they could build a tower to the sky, they could climb up and be like gods themselves. Mankind, in just a few generations since the flood, had once again lost sight of God and become prideful and self-sufficient.

   God put an end to their plans by confusing their language so they would have to spread out over the earth.

We may be tempted to think we no longer need to trust and obey God. We may believe our money, family, education, or resources are all we need to guarantee our security. Scripture tells us otherwise (Ps. 37:13, Prov. 14:12, Eph. 2:8-10).

Worksheet 5, Section Five, Part 4

Terms to Know and Spell

sheet erosion — the erosion of material by water flowing over land as a widespread mass instead of in definite channels or rills.

water gap — a pass or gorge carved by flowing water through a mountain, ridge, or plateau when there appears to have been an easier route around the barrier.

sinkhole — a hole formed in soluble rock (especially limestone) by the action of water, serving to conduct surface water underground.

Short Answer Questions

1. They can't. Radioisotope dating methods do not provide reliable results.
2. rodents
3. Cedar Mountain is an erosional remnant left after the Flood.
4. Ravens
5. The San Francisco Peaks

Discussion Questions

1. Discussion should include that with the current rate of encroachment (1 foot per year) of the surrounding forest into the meadows, there would not be any meadowland left if the earth were millions of years old.
2. Limestone is never far beneath the surface of the soil, as it is the top layer of rock capping most of the Grand Canyon. Water percolates through the relatively porous limestone, dissolving it, and creating underground caverns. If these caverns collapse, an indentation or sinkhole is formed.

Worksheet 6, Section Five, Part 5

Terms to Know and Spell

symbiotic relationship — the close association between two or more organisms of different species, often but not necessarily benefiting each member.

folds/folding — an occurrence where originally flat sedimentary layers are bent or curved.

underfit stream — a stream that appears too small to have eroded the valley or canyon in which it flows and too small to remove the associated rock debris.

Short Answer Questions

1. Many scientists believe the Anasazi left the area to find a more favorable climate.
2. rim, canyon floor
3. evaporation
4. 1908
5. small pockets of unburned areas

Discussion Questions

1. The top of the Dox Sandstone has been eroded almost table-top flat. This is consistent with Flood catastrophic currents but is unexplainable by the evolutionary model of slow erosion over millions of years, as slow erosional processes do not form large flat surfaces.
2. The folds are found throughout the Canyon and in several of the layers. This indicates that the layers were folded while still soft and pliable, which likely took place during the receding of the Flood.

Worksheet 7, Section Six - Seven

Terms to Know and Spell

sapping — the process in which groundwater exits a bank or hillside laterally in the form of a seep or spring, eroding soil from the slope and often causing the collapse of material above.

weathering — the process of breaking down rocks, soils, and their minerals through direct or indirect contact with the atmosphere and moisture, which, contrary to erosion, occurs without movement of the particles.

circular reasoning — the logical fallacy in which a premise presupposes the conclusion in some way, yet provides no reason at all to believe its conclusion.

Short Answer Questions

1. Mount St. Helens, Channeled Scabland
2. operational science
3. calcium carbonate, silica, iron oxide
4. time
5. a rapid of flow of water in such huge quantities would have washed the rock debris away

Discussion Questions

1. The "breached dam" model proposes that after the Flood, two or three very large lakes were ponded east and north of the Kaibab Plateau. After some period of time, probably a few hundred years or less, these lakes breached through their natural dams and carved the Grand Canyon catastrophically.

   The second hypothesis proposes that the Grand Canyon was carved late in the Flood. Following massive sheet erosion of the area during the Flood's recession, the water currents channelized as more and more land became exposed. It was during this channelized phase of the Flood that a westerly moving current would have carved the Grand Canyon through the uplifted Kaibab Plateau.

2. Discussion should include that in the evolutionist scenario of millions of years, we should see evidence of tremendous erosion at the boundaries between the layers where rivers, streams, and other erosional agents removed material.

   The creationist view is that the sedimentary layers were all deposited during a single catastrophic event of global proportions about 4,500 years ago. The expected evidence for this explanation should show continuous deposition of sediments, one layer after the other, with regular flat boundaries showing very little, if any, erosion between them.

Worksheet 8, Section Eight

Terms to Know and Spell

phyla — (plural of phylum) a taxonomic rank at the level below Kingdom and above Class in biological classification, especially of animals.

kind — a term used in the Bible that is approximately equivalent to the modern classifications for the family or genus of animals or plants.

Short Answer Questions

1. Cambrian Explosion
2. transitional fossils
3. either "creation model" or "biblical Flood" would be acceptable answers
4. nautiloids

Discussion Questions

1. If fossilization occurred as a slow process, you would expect to find coral reefs and beds of clams all in their "natural" position. But instead clams are found orientated in every direction with their shells still closed, indicating they were still alive but moved when buried, and corals are broken into small pieces.
2. The fossil bed occupies an area of at least 5,700 square miles and, based on exposed areas, contains an average of one fossil per square yard. The fossils and their directional orientation indicate that these nautiloids experienced a mass kill event and were deposited in massive underwater landslides moving in a westerly direction.

Worksheet 9, Section Nine

Terms to Know and Spell

riparian — the ecosystem of plants and animals living along a water source that are dependent on that water source.

biological province — an area of land, less extensive than a region, having a characteristic plant and animal population.

region (ecological) — a part of the earth characterized by distinctive animal or plant life.

ecosystem — a system that includes all living organisms (in an area as well as its physical environment) functioning together as a unit.

Short Answer Questions

1. climate
2. Mojave, Sonoran, Great Basin
3. cactus, succulent
4. food, nesting
5. the dipper or "water ouzel"

Discussion Questions

1. Prior to the dam, the Colorado River was a deep red-brown color due to the enormous amount of sediment it transported downstream The river corridor was subject to seasonal flooding, which scoured out the riparian vegetation growing along the riverbank, stirred up the bottom sediments, and renewed beach areas. The colder water temperatures have caused the extinction of nearly half of the native fish, with most of the remainder considered endangered – but it now supports an introduced trout population.

Worksheet 10, Section Ten

Short Answer Questions

1. desert culture
2. Pueblo culture, Anasazi
3. 1869
4. Theodore Roosevelt

Discussion Questions

1. Discussions will vary.

Awesome Science Explore Yellowstone DVD Worksheet Answer Key

Worksheet 1
About Yellowstone National Park

1. Geysers, mud pots
2. Super volcanoes, eruptions
3. Geologic, plates
4. Flood
5. Sedimentary, basement, volcanic rhyolite, tuff
6. Caldera
7. Lava chamber
8. Sin-cursed

Discussion Questions

1. cemented volcanic ash
2. Plant life above the ground may be destroyed in a fire, but many plants will come back after the fire.
3. Yes

Worksheet 2
Geologic Column

1. Precambrian, Cenozoic
2. Sedimentary, catastrophic
3. Fossils, layer
4. Precambrian
5. Flood, Creation week
6. Erosion, fossilization
7. Microbes
8. Paleozoic
9. Vegetation, log mats
10. Sediments, buoyancy
11. Biblical worldview

Sequence to create coal:

3, 6, 7, 4, 2, 1, 5

Worksheet 3
Discussion Questions

1. Answers will vary. These examples of biblical text reveal geologic details and history that we can see in the world around us today. The Bible is not a collection of stories, but an actual history book of the world – including geology.
2. The traditional belief from secular scientists is that the geologic column represents long ages of time, billions of years, starting with the Precambrian and ending in the Cenozoic, being laid down slowly and gradually without a global catastrophe. This sequence is never complete in any one location, so secular scientists have to combine locations to get their full view of the column.
3. 1,600 years
4. Since many of the layers are sedimentary, the Flood was the catastrophic historical event that was the mechanism for quickly laying down each layer. The conditions necessary for fossilization didn't occur until the Flood came.
5. When the flood moved inland, it wiped out most of the vegetation. Land animals would have

escaped to higher ground, or by virtue of being on higher ground, would have a higher burial in the fossil layers.

6. Mesozoic
7. It is suspected that mammals survived the longest. Many of them could have survived a bit longer on higher ground or on the floating log mats. Many mammals also float when dead, preventing them from being rapidly buried by sediment, and fossilized.
8. Since there is not a full sequence of the column in any one location, this is consistent with the Flood account because water and currents will carry sand, silt, and mud to different places around the earth. Secular scientists say early layers do not have fossils because life was too "simple" at the beginning, and only microbes had evolved. But the biblical explanation is much stronger for several reasons. First of all, these "simple" life forms are not simple at all! And further, the conditions necessary for fossilization didn't occur until the Flood came, and the sequence of the Flood is reflected in the sequence of fossil layers that we see today – for example, Paleozoic layer shows fossils on the sea floor creatures, which would have been the first to be impacted by the Great Flood because it started with the "fountains of the great deep opening up."
9. Dead animals on the sea floor are quickly consumed by scavengers, and their bones decay away – the same process is true on land.
10. Cenozoic

Worksheet 4
The Geysers

1. True
2. True
3. False
4. False
5. True
6. True
7. False
8. True
9. False
10. True
11. True
12. True
13. True

Worksheet 5
Yellow River and Hayden Valley

1. Fumaroles
2. Yellowstone Lake
3. Buffalo
4. Hayden Valley, Ice Age
5. Glacial, shoreline

Discussion Questions

1. Hayden Valley
2. It has cut its way through it.
3. The river cutting through the soft tuff and rhyolite.

Worksheet 6
Yellowstone Falls

1. Grand Canyon of Yellowstone
2. Tens of thousands of years; ice dam
3. Fast-moving
4. Missoula Flood

Discussion Questions

1. The steepness of the canyon walls helps us realize there has been a recent cutting of the canyon. Based on the current rate of erosion, long periods of time would have eroded the canyon much more than it is now and the canyon walls would have been much more gradual in their slope.
2. We know this is not formed by glaciation because it does not exhibit U-shaped features, like at Yosemite National Park. A large glacier, as it moves and cuts the valley underneath it, will shape that valley with rounded corners at the bottom. When the ice retreats, the valley has a U-shaped bottom. A valley cut by a river tends to have a V-shaped valley, with steep sides.
3. A large glacier, as it moves and cuts the valley underneath it, will shape that valley with

rounded corners at the bottom. When the ice retreats, the valley has a U-shaped bottom.

4. Answers will vary. The discussion should be focused on taking the evidence we see in the world around us and how it was formed rather than making assumptions based on secular dating processes or theories.

Bonus Activity:

Answers will vary. This can be used as an essay, a discussion, or further research.

Worksheet 7
Mud Volcano

1. False
2. False
3. True
4. False
5. True

Worksheet 8
Yellowstone Lake

1. Mount St. Helens, valley
2. Dormant
3. West coast
4. Catastrophe, water, magma, equalize
5. Answers will vary.

Worksheet 9
Old Faithful

1. 95 minutes
2. 1959, 1983

The Mammoth Hot Springs

7, 9, 1, 6, 3, 4, 8, 2, 5

Worksheet 10
Petrified Forest

1. True
2. True
3. True
4. False
5. False
6. True
7. False
8. True
9. True
10. False
11. True
12. True

Bonus activity

Answers will vary. The importance of the activity is looking to the Bible for answers for the world around us, for history, geology, and more than just faith questions.

Worksheet 11
What Is Petrification?

1. False
2. True
3. True
4. False
5. True
6. True
7. False
8. True
9. True
10. True

Awesome Science Explore Meteor Crater and Petrified Forest
DVD Worksheet Answer Keys

Worksheet 1
The Grand Staircase and Noah's Flood

1. remnant
2. Grand Staircase, Bryce
3. geologic
4. Genesis 6
5. fountains, water
6. mountain ranges
7. receded, continental
8. Kaibab
9. volcanoes
10. Chinle Formation
11. ash, underwater

Discussion Questions

1. A remnant of a vast, dry lake bed
2. Valleys are eroded, mountains rise, and lakes are formed

Worksheet 2
About the Petrified Forest

1. True
2. True
3. False
4. True
5. False
6. True
7. True
8. True
9. True
10. True
11. True
12. True
13. True
14. False
15. True
16. True
17. True

Discussion Question

1. Everything happened quickly, and all the trees were covered in water and sediment, causing them to fossilize.

Worksheet 3
Other Petrified Examples

1. catastrophic
2. landslide
3. tidal wave
4. Spirit Lake
5. peat
6. silica
7. erosion

What About the Teepees?

1. water erosion
2. monsoon
3. layers
4. deposited
5. Noah's Flood
6. underwater
7. sediments
8. uniformity

Worksheet 4
Rapid Layer Deposition and Erosion

1. processes
2. debris
3. hundreds
4. crater
5. melted
6. earthen dam
7. erosional
8. Grand Canyon
9. features

10. airborne
11. strata
12. alternating
13. independent
14. periods
15. ash layers
16. formations

Discussion Questions

1. Answers will vary.
2. Answers will vary. Erosion is a process of tearing away soil or stone by wind, water, or other natural means. It can completely change the landscape depending on the length of time and amount of force.

Worksheet 5
Rapid Layer Deposition and Erosion

1. Dinosaurs were buried quickly during the Flood.
2. It is not an ancient forest floor, but strata in sediment laid down repeatedly by water. This shows quick formation.

Worksheet 6
About Meteor Crater

1. model
2. meteorite
3. asteroid
4. heat
5. Shoemaker
6. Yucatan Peninsula
7. dust storm
8. starved
9. dust layers
10. acid rain

Discussion Questions

1. It is a fragment of space matter often made of iron and nickel.
2. They are not buried in dust, but in sedimentary rock layers.

Worksheet 7
Meteor Crater Early History

1. True
2. False
3. True
4. True
5. True
6. False
7. True
8. False
9. False

More Details on the Crater

1. True
2. True
3. False
4. False
5. True
6. True
7. False
8. True

Shoemaker's Theory

1. hypothesized
2. anomaly
3. craters
4. meteorites
5. evidence
6. sea floor
7. Chesapeake
8. moon
9. ashes

Worksheet 8
Dinosaur Extinction Because of Meteor Impact?

1. global flood
2. water
3. carcasses
4. quickly

5. Bible
6. quick burial
7. land-dwelling
8. perished
9. dust storm
10. population
11. numbers
12. tidal waves

Awesome Science Explore Yosemite/Zion DVD Worksheet Answer Key

Worksheet 1
About Yosemite National Park

1. Granite
2. Glory
3. Nature, order
4. Rhode Island, alpine
5. California, inspiration
6. Yosemite Valley, El Capitan
7. Deep valley
8. Millions
9. Formation, glaciers
10. Biblical

Discussion Questions

1. Answers can vary. For example: Secular scientists think that it took millions of years, and processes like erosion from the Merced River and multiple ice ages over long periods of time.
2. Answers can vary. For example: The biblical model of one major Ice Age after the Great Flood supports the evidence we see in the world today versus many ice ages over time, an idea scientists have dozens of ideas on how they may or could have formed.

Worksheet 2
Evidence for Quick Formation of Granite

1. True
2. False
3. True
4. True
5. False
6. True
7. True

Discussion Questions

1. Answers can vary, but should focus on helping to determine the age of the Earth. For example: By noting the detailed genealogies from the Bible, you can form a timeline of the age of the Earth back to the point of creation, and they show an unbroken timeline of people – just like if you were tracing back your own family tree.
2. Because the story of evolution is so widely believed, it seems like there is a lot of authority in what secular scientists say. Also science says they have "proven" the things they say, though in some cases that is impossible. If you assume everything they tell you is correct or has been proven by some method (not telling you how flawed that method might be), it seems to be evidence versus just what the Bible says is true, unless you realize the Bible is actually the history book of the universe.

Worksheet 3
The Receding Floodwaters: Cutting of Yosemite

1. Cliffs
2. Glacier Point
3. Mountains, Merced
4. Erosion, water
5. Catastrophe
6. Muir, hypothesize
7. Global
8. Continents, chains

Worksheet 4
Evidence for Quick Formation of One Ice Age

1. True
2. True
3. True

4. False
5. False
6. True
7. True
8. False
9. True
10. True

Worksheet 5
Fill in the Blanks

1. Silver Lake
2. Systems
3. Precipitation
4. Large-scale
5. Summers, sunlight
6. Equalize, oceans
7. Ice
8. Ice Age, flood
9. Accumulation
10. Model
11. Meadows, winter
12. Snowfall

Evidence for Quick Melting of the Ice Age

1,3,2,4

Worksheet 6
About Zion National Park

1. Utah
2. Grand Staircase, Arizona
3. National Park
4. Bottom, habitation
5. Cathedrals
6. References, patriarchs, angels
7. Crossbeds
8. Flowing
9. Sandstone
10. Zion
11. Challenges, understanding

Worksheet 7
Formation of the Layers

1. True
2. False
3. True
4. True
5. True
6. True
7. True
8. False
9. True

Worksheet 8
Cutting of Zion Canyon

1. Virgin River
2. Gradually
3. Features, cut
4. Canyon, gravity, forces
5. V-shaped, younger
6. Arches
7. Collapsed, hanging
8. Directions, higher
9. Intersections
10. Water
11. Small, remnant, receding
12. Big Bend
13. Obstruction

Worksheet 13
Discussion questions:

1. The Virgin River flows too gradually to have caused this much erosion. The river doesn't even reach the canyon walls. Other features in the canyon make clear this place was cut by a lot of water over a very short period of time, possibly just days or hours. Arches and hanging valleys are two other examples of how the river could not have carved the canyon over a long period of time.

2. If the canyon was formed over a long period, gravity, erosion, and other natural forces would have caused these steep walls to collapse into the canyon in various places. As a result, this valley would have been V-shaped with lots of debris.
3. There would be more debris, but in truth, there is very little debris on the canyon floor

Awesome Science Explore Mount St. Helens DVD Worksheet Answer Keys

Worksheet 1
The Grand Staircase and Noah's Flood

1. high, Cascade
2. British Columbia, Flood
3. tectonic
4. active, dormant
5. Mount Rainer
6. Washington, virgin
7. Spirit, Toutle
8. earthquakes
9. sixty
10. steam, magma

Worksheet 2
About Mount St. Helens

1. True
2. False
3. True
4. True
5. False
6. False
7. True
8. True
9. False
10. False
11. True
12. False
13. True
14. True
15. False
16. False
17. True

Discussion questions:

1. Answers will vary.
2. The earth trembles and the mountians smoke.
3. Small. Larger.

Worksheet 3
The Grand Staircase and Noah's Flood

1. peaks, extinct
2. Mazamma, Pacific
3. Oregon, hot
4. south, formations
5. lava, Ape
6. "Bigfoot"
7. foresters
8. flow
9. Kilauea

Order of Events:

6, 4, 5, 1, 2, 3

Discussion questions:

1. 2,000 years old. A mud flow also helped shape it.
2. Because the mountain usually produced much thicker lava.
3. If it had been around for millions of years, it would have eroded or collapsed by now. Answers will vary.

Worksheet 4
Rapid Strata Formation

1. False
2. True

3. True
4. False
5. True
6. False
7. True
8. True
9. True
10. False
11. True
12. True
13. True

Discussion questions:

1. Answers will vary.
2. These assumptions are simply not true because the layers were formed in a very short time.
3. It is the opposite of uniformitarianism. Things can change
4. They don't want to believe in God or in judgement.

Worksheet 5

1. geology
2. lamina, pyroclastic
3. homogenized
4. micro-thin lamination, fine
5. Secular, seconds
6. canyons
7. sedimentary, particles, elevation
8. evidences, water
9. receding, landscape

Discussion questions:

1. They assume that the earth is millions of years old and things changed slowly.
2. The evidence showed there could be massive changes in a short period of time.
3. Canyons don't and millions of years to form but actually only days or weeks.

Worksheet 6
True of False:

1. True
2. True
3. True
4. False
5. False
6. True

Fill in the Blanks:

1. ideas, Kaibab Plateau
2. dam, eroded
3. receding, trapping
4. creationists, gaps

Worksheet 7

1. True
2. False
3. True
4. False
5. False
6. False
7. True
8. True
9. False
10. False

Fill in the Blank:

1. waterlogged
2. root balls, upright
3. soaked, sink
4. Species
5. resin
6. absorption
7. bark
8. living
9. interpret
10. mystery
11. Spirit Lake, mapped

12. sonar

Fill in the Blank:

1. rings, burrowing
2. Yellowstone, park sign
3. uniformitarian
4. great deep
5. subterranean, fissures
6. log mats
7. ash
8. pressure, chemical
9. long ages, petrify

Worksheet 8
True of False:

1. True
2. False
3. False
4. True
5. True
6. False

Order of Events:

2, 5, 4, 1, 3

Discussion question:

It could only happen rapidly, because there are no roots.

Worksheet 9
Biological Recovery

1. toxic
2. organic, oxygen
3. Legionella
4. bacteria, plankton
5. hundreds, design
6. sections, biological recovery
7. spring, hibernation
8. pocket
9. blast zone
10. scat
11. nitrogen
12. Lupine
13. alders
14. algae
15. side, reintroducing
16. nutrients

Discussion questions:

1. Answers will vary.
2. God has created a world that recovers quickly from catastrophe.
3. Responses will vary.
4. Answers will vary.

Awesome Science Explore John Day Fossil Beds DVD Worksheet Answer Keys

Worksheet 1
About John Day Fossil Beds

1. Oregon
2. coastline, basalt
3. fissures
4. lava flows
5. thousands, volcanic
6. fossils, plants
7. monument, three
8. expedition
9. John Day

Discussion questions:

1. Answers will vary.
2. Answers will vary.

Worksheet 2
The Formations

1. True
2. True
3. True

4. False
5. True
6. True
7. True

Fill in the blanks:

1. during, balance
2. pillows
3. evolutionary
4. quickness
5. magma
6. post-flood, thousands
7. local
8. lava beds
9. short
10. slumping

Discussion questions:

1. They are spread over such a wide area.
2. Yes

Worksheet 3
Erosion Of The Formations

1. excavation
2. properties, Mount St.Helens, canyons
3. movements, elevation
4. Pacific Ocean
5. ocean basins, sheet erosion, shelves
6. gaps
7. erosional

Worksheet 4
The Palisades

1. conglomerates, secular, lahars
2. high desert, jungle
3. leaves, avocado
4. bedding, sorted
5. volcanic mudflow
6. eruptions, particles, ash

Petrified Logs:

1. True
2. True
3. True
4. True
5. True
6. True
7. False
8. False
9. True
10. True
11. True
12. True
13. True
14. False
15. True

Worksheet 5
The Painted Hills

1. paleontologists, rainfall
2. floodplain, drying
3. flora, destruction, evidence
4. root balls, coal
5. permineralized
6. vegetation, pre-Flood

Discussion questions:

1. They would have escaped a local flood.
2. The leaves were flat indicating they were submerged in water and buried quickly.

Worksheet 6
The Cretaceous Gable Creek Conglomerate

1. True
2. False
3. True
4. True
5. True
6. True

7. False
8. True
9. True

Picture Gorge

1. True
2. True
3. True
4. True
5. False
6. True
7. True
8. True
9. True

Discussion questions:

1. The canyon carved the earth, which formed the passage for the river.
2. The first part

Worksheet 7
Sheep Rock

1. Colors, formation
2. disjointed, localized, violent
3. cap
4. connected
5. sediments, Mascall
6. diatoms

Discussion questions:

1. Erosion would have taken place.
2. Tortoises and saber-toothed tigers
3. It can be much hotter in the summer afternoons.

Worksheet 8
Mammals of John Day Fossil Beds

1. True
2. False
3. True
4. True
5. False
6. True

Discussion questions:

1. No. Dog kind.
2. Varieties or species within animal kinds can develop as a result of losing information. The opposite of evolution.
3. The Flood currents would have brought diverse mammal fossils and layers.

Worksheet 9
Mammals of John Day Fossil Beds

1. Pediments are gently inclined surfaces carved into bedrock by erosion.
2. Yes. The current moved down the valley, eroding what we see today, producing wide-valley pediments.

Your Guide to Yellowstone and Grand Teton — Worksheet Answer Keys

Worksheet 1, Foreword – Introduction

Terms to Know

caldera — a depression formed when a volcanic eruption empties the underground reservoir of magma, which is followed by the collapse of the roof of the magma chamber.

plateau — a landmass with a relatively flat summit area that is elevated above adjacent terrain and bounded on at least one side by a steep slope.

magma — naturally occurring molten rock material, generated within the earth and capable of moving through fractures in solid rock and cooling there or flowing onto the surface.

evolution — the belief that all known organisms are related by a common ancestor, having evolved over billions of years by a process in which inherited traits becomes more or less common in a population over successive generations.

Short Answer Questions

1. 8,000 feet
2. The Grand Tetons
3. 18 million; 28,000
4. 10,000

Discussion Questions

1. Discussion should include that without God and without a true knowledge and understanding of Scripture, which provides the true history of the world, man is doomed to repeat the same mistakes over and over again.
2. The student should display an understanding that just as Noah and his family had to go through a doorway into the Ark to be saved, so does mankind today. Because of God's great love for us He provided another door for us to enter through, that door being Jesus Christ. Additional Scripture: (John 14:6) Jesus answered, "I am the way and the truth and the life. No one comes to the Father except through me."

Worksheet 2, Sections One – Four

Terms to Know

uniformitarianism — the philosophy that assumes the natural processes operating in the past were the same as those observed operating in the present, often summarized in the statement, "The present is the key to the past."

creation — the concept that all humanity, life, and the earth (the universe as a whole) was created by God.

geologic time — a term used by evolutionary geologists to describe the time they believe occurred during earth's history.

sedimentary rocks — a layered rock formed by the accumulation and consolidation of sediments.

Noah's flood — the global Flood described in the Bible that came as a result of God's judgment of sin.

worldview — the framework through which an individual interprets the world and interacts in it.

Short Answer Questions

1. The rapid burial and sedimentation that occurred during Noah's flood
2. God's Word, the Bible
3. Out of nothing (ex nihilo)
4. One and a half

Discussion Questions

1. Discussion should include points highlighted on page 13; leaving places in better condition than we find them, recycling, using designated trails, not picking wildflowers, and avoiding feeding the wildlife. In addition, we want to follow any other guidelines posted at national parks we may visit.
2. Explanations will vary but should demonstrate a basic understanding that Mr. Kepler was stating his belief that God is the Creator of all things and that science should be interpreted through an understanding of God through His Word.
3. Students should display a basic understanding that living systems require all multiple component parts to be present and functioning together from the start; one innovation cannot wait for another to come along and make it functional. Additionally, almost all mutations result in the loss of information and function.

Worksheet 3, Section Five, Part 1

Terms to Know

thermophiles — a "heat-loving" micro-organism that lives in or around hot springs and hot pools that include bacteria, viruses, archaea, and eukarya.

rhyolite — a high-silica volcanic rock that is usually too thick to flow.

supervolcano — a very large volcano capable of producing an eruption of more than 240 cubic miles of material.

radiometric dating — a technique used to date metamorphic and igneous rocks based on the assumed decay rate of the specific elements and the resulting ration of parent and daughter elements found in the rocks today.

Short Answer Questions

1. Volcanic caldera.
2. Geothermal activity.
3. mud pots, fumaroles, hot springs, geysers.

4. Vegetation around hot springs is high in fluoride and laced with silica, which wears down their teeth prematurely, causing them to die of starvation.
5. An earthquake caused changes to Old Faithful's subterranean "plumbing."
6. The first eruption in Yellowstone likely occurred during or shortly after the receding of the Flood.

Discussion Questions

1. Discussions will vary.
2. http://www.yellowstonepark.com/scenic-drive-continental-divide-tour-colorado-to-yellowstone/

Worksheet 4, Section Five, Part 2

Terms to Know

quartzite — a metamorphic rock consisting mainly of quartz, formed by recrystalization of sandstone when under pressure from deposits above it.

breccia — a coarse-grained rock, composed of angular broken rock fragments held together by a mineral cement or a fine-grained matrix of small rock particles.

hotspot — a volcanic region that is thought to be the surface expression of a rising plume of hot mantle material.

layer — rock or soil unit with internally consistent characteristics that distinguish it from those above and below.

Short Answer Questions

1. The long-distance transport of these rocks is readily explained by rapid currents retreating off the rising continents at the end of the Flood.
2. Moose.
3. Geothermal activity.
4. A global flood.
5. The majority of the bark missing from trees is caused by bison (and sometimes elk) using the trees as scratching posts. Bears will claw or bite the trees.
6. six, eight

Discussion Questions

1. Discussions will vary.

Worksheet 5, Section Five, Part 3

Terms to Know

creationist — a person who believes that all humanity, life, and the earth (the universe as a whole) was created by God.

igneous rocks — rocks formed when molten rock (magma) cools and solidifies, either below the surface as intrusive (plutonic) rocks; or at or near the surface as extrusive (volcanic) rocks.

andesite — a dark colored, fine-grained extrusive igneous rock with visible dark and light minerals such as sodium-calcium feldspars and hornblende.

basalt — an igneous rock formed from molten lava, which normally flows on the surface.

hoodoos — a pillar of rock formed by erosion, often capped by harder, more resistant rock.

erosion — the displacement of solids (soil, mud, rock, and other particles) by the action of wind, water, or ice.

Short Answer Questions

1. The colors of the rhyolite lava walls of the Grand Canyon of the Yellowstone are caused by hydrothermal effects on iron.
2. It is estimated that ice around Mount Washburn was as much as 3,000 feet thick during the Ice Age. As the main ice sheet moved south, the huge granite boulders found on Mount Washburn were transported "uphill" from the Lamar Valley.
3. The soil in this area is derived from andesite and is rich in minerals.
4. Basalt columns are formed when thick lava flows cool and contract, creating shrinkage cracks resulting in predominantly polygonal columns.

Discussion Questions

1. Lodgepole pines produce two kind types of cones, one of which will release seeds only after reaching a temperature of at least 113 F. (See pgs. 35 and 75). For further discussion points visit the national park service website at www.nps.gov and search under "fire ecology."
2. Discussion should include that the Grand Canyon is narrow, steep-walled, and V-shaped,

indicating water carved it not long ago. Evolutionists believe there were multiple ice ages, each lasting 100,000 years, which would have carved huge U-shaped canyons.

Worksheet 6, Section Five, Part 4

Terms to Know

fossil — the mineralized or otherwise preserved remains or traces (such as footprints) of an animal or plant preserved primarily in sedimentary rock.

travertine — a coating or buildup of limestone formed by the evaporation of water.

Short Answer Questions

1. Rapidly
2. Because of the quantity and variety of large animals living there, including; bison, pronghorn, coyotes, beaver, mule deer, fox, elk, wolves, and grizzly bears.
3. Elk provide about 90 percent of a wolves' winter diet and are plentiful in the Lamar Valley.
4. Thermophiles
5. Specially designed enzymes enable the thermophile's DNA to replicate in extremely high temperatures.
6. Water and steam

Discussion Questions

1. Present-day processes do not explain how a mountain could slide 30 miles down a very flat slope. But the powerful upward movement of the earth's crust late in the Genesis Flood, while the area was still underwater and slippery, provides a plausible mechanism. Strong earthquakes caused by vertical uplift would have likely dislodged the large block of carbonate rock layers, freeing it to move very rapidly.
2. Discussion should include the creationist view that a catastrophic deposit of plants and animals originated from different locations during a global flood versus the uniformitarian explanation that the marine fossils originated from a 70-140 million year old shallow inland sea. (Also see Genesis 7:17-24.)

Worksheet 7, Section Six, Part 1

Terms to Know

metamorphic rock — the result of the transformation of a pre-existing rock by heat (but below the melting point) and extreme pressure, causing profound physical and/or chemical change.

brachiopod — a two-shelled marine animal resembling a clam, though generally with its two shells being of different sizes.

gneiss — a metamorphic rock with bands of dark and light minerals formed under pressure of deep burial, causing the minerals to form bands.

basement rock — the rocks below a sedimentary platform or cover, or more generally any rock below sedimentary layers that are metamorphic or igneous in origin.

moraine — a mound or ridge of mixed rocks, sands, and silts deposited by direct action of flacire ice and occurring as a lateral, terminal, or ground moraine.

dike — a wall-like body of igneous rock that has been intruded into the rock, cutting across the structure of the surrounding rock.

terrace — any long, narrow, relatively level, or gently inclined surface bounded along one edge by a steeper descending slope and along the other by a steeper ascending slope.

Short Answer Questions

1. 1894
2. When hunting together with badgers, coyotes catch and share about one-third more rodents than when hunting alone
3. Granite and gneiss
4. Beavers

Discussion Questions

1. There are no signs of erosion within or at the boundaries between the layers, indicating the whole sequence of layers was deposited quickly. In addition, the sequence of rock in these layers is the same sequence covering the basement rocks across much of North America. Present-day geological processes cannot explain such

extensive formations.

Worksheet 8, Section Six, Part 2

Terms to Know

fault scarp — a topographic surface expression of faulting attributed to the displacement of the land surface by movement along a fault.

glacial outwash — sand and gravel deposited by melt water from a glacier.

biodiversity — the variety of life within an ecosystem or community.

contact — the boundary surface between two types of ages or rocks; (contacts are either "conformable" if they represent no gap of time between them, or an "unconformable" if they are thought o represent a gap of time or erosion between the rocks).

Short Answer Questions

1. Glaciers
2. Clouds, glaciers, streams, and lakes are all part of the water cycle.
3. 70 percent
4. Ice caps and glaciers

Discussion Questions

1. Discussions will vary.

Worksheet 9, Section Six, Part 3

Terms to Know

limestone — a sedimentary rock composed largely of calcium carbonate.

sandstone — a sedimentary rock composed mainly of sand-sized mineral or rock grains cemented together.

shale — a fine-grained sedimentary rock whose original constituents were clay, mud, and/or fine silt.

Short Answer Questions

1. A diversity of plants, animals, and microorganisms increases the health and productivity in an ecosystem.
2. These are called percussion marks and indicate a very turbulent flow as the boulders were carried here by the Flood.
3. Most pronghorn migrate 100 miles southeast to the Green River Basin.
4. Pronghorn

Discussion Questions

1. Similar sedimentary layers cover other basement rocks in the Rocky Mountains, and some layers extend across most of North America. Believing that today's slow processes of erosion and deposition account for such continental-scale sedimentation is unrealistic. This widespread deposition makes sense in the context of a global flood.

   In addition, the contacts between the layers do not show any uneven erosional features, as would be expected if these layers were laid down over millions of years. Instead, the layers look like they were rapidly laid down one after the other in quick succession.
2. Pronghorn are especially designed for life on the open plains. Able to sprint at over 60 mph, they are the fastest mammals on the continent. Since their body hair is hollow and can be lifted or flattened at will, pronghorn are able to adjust to temperature extremes. Standing their hair erect allows air to cool their skin, whereas laying their hair down flat retains heat. Pronghorn, often mistakenly called antelope, have horns made of keratin like cattle, but shed their "pronged" horns like deer, placing them in a unique category between the cattle and deer kind.

Worksheet 10, Section Seven

Terms to Know

empirical (or observable) science — a method of careful investigation, repeatable experiments, and a skeptical attitude that insists all theories be carefully tested.

historical (or forensic) science — a set of beliefs that uses scientific terms and ideas to explain events that happened in the unobservable, untestable past.

tectonics — a branch of geology dealing with the broad features of the outer part of the Earth, especially in relation to its vertical and horizontal motions.

formation — a body of rock layers consisting predominantly of a certain type or combination of types of rock.

naturalism — any of several philosophical beliefs that the natural world is the whole of reality and does not distinguish the supernatural from nature.

Short Answer Questions

1. Worldview
2. Geology and dating methods are the foundations of evolution.
3. Slow, gradual processes
4. Marine fossils

Discussion Questions

1. The Tetons and Jackson Hole display a tremendous amount of past vertical movement, called tectonics (see page 150). Such tectonics, in which part of the upper crust moves up and part sinks down, are seen in many parts of the world. The greatest change in Wyoming is between the 14,000-foot high granitic Wind River Mountains, just southeast of the Teton Range, and the granitic basement rocks found 31,000 feet below sea level in Hanna Basin. This amounts to a net vertical change in the upper crust of 45,000 feet in the state of Wyoming alone! Critics of the global Flood commonly point out that the Flood water could not cover Mount Everest. These critics do not understand that Mount Everest and the Himalayan Mountains, which are made up of sedimentary layers containing marine fossils, rose out of the Flood water. If Wyoming can have a change in the upper crust of 45,000 feet, it seems to be no problem for the Himalayan Mountains to be pushed up 30,000 feet.
2. Discussions will vary.

Worksheet 11, Section Eight

Terms to Know

Channeled Scabland — an area of eastern Washington characterized by vertical walled canyons and flat bottoms caused by the Lake Missoula flood.

carbon dating — a radiometric dating method that uses the naturally occurring isotope carbon-14 to determine the age of carbonaceous materials.

geologic column — the column of sedimentary rock that graphically represents the sequence of rock formations for a given locality or region (the entire global geologic column shown in most textbooks is based on a compilation of all the local columns according to the uniformitarian subdivisions of geologic time and is not found anywhere on earth as a continuous sequence).

parent element — the isotope or precursor atom from which a daughter element is derived during radioactive decay (e.g., potassium is the parent element of argon gas).

daughter element — the isotope resulting from the radioactive decay of a parent isotope (e.g., argon gas is the daughter element of potassium).

paleontologist — a scientist who studies past life forms on earth through the examination of plant and animal fossils.

Short Answer Questions

1. Ice, tectonic uplift
2. Contacts
3. Terminal, lateral
4. Radioactive decay

Discussion Questions

1. Discussion should include the importance of understanding the worldview and basic assumptions made before the known facts were interpreted. Understanding what those assumptions are is important. If we are not aware of the assumptions that are being used, we can easily be deceived. We should always start with the Bible, the ultimate source of truth.

Worksheet 12, Sections Nine – Ten

Terms to Know

transitional fossil — the fossilized remains of a life form that illustrates an evolutionary transition; a human construct that vividly represents a particular evolutionary stage, as recognized in hindsight.

biotic communities — or "natural communities," often used by ecologists to classify different habitats and their plant and animal associations.

mutation — a permanent change in the DNA sequence that makes up a gene, leading to changes that may be helpful or harmful to the organism.

theory — a concept or proposition developed and better substantiated than a hypothesis but not so conclusively proven as to be accepted as a law.

Short Answer Questions

1. Buried quickly
2. Minerals
3. Lodgepole pine
4. Elk
5. Seven
6. Bison

Discussion Questions

1. Discussion should include that fossils are often found in huge "fossil graveyards," reflecting widespread and violent conditions, unlike anything happening today. A global flood would explain such widespread and violent conditions. The Flood would result in quick, deep burial and extreme pressures causing mineral-charged water to flow into the organism.
2. Even though instituted with good intentions, the fire suppression programs helped to create an environment that led to the largest fire-fighting effort in the history of the nation, burning 36 percent of Yellowstone. The summer of 1988 was the driest on record in Yellowstone. The gigantic fuel stores of downed trees and forest litter, the record drought, plus double the usual number of lightning storms combined to make "the perfect fire-storm."

Awesome Science Explore Glacier National Park DVD Worksheet Answer Keys

Worksheet 1
About Glacier National Park

1. Montana; Canada
2. stomatolites
3. ice ages
4. global Flood
5. Salish; Flathead; Shoshone; Cheyenne
6. Crown of the Continent
7. President Taft
8. Lake McDonald
9. pollution
10. Proterozoic
11. genealogies

Secular Ideas for Mountain Formation

1. False
2. True
3. True
4. False
5. False

Worksheet 2
The Ice Age

1. U-shaped valleys
2. Moraines
3. uniformitarianism
4. snowball earth
5. abundant moisture; mild winters; cool summers
6. oceans
7. sub-zero
8. ash; aerosols
9. computer model
10. observe

The Biblical Answer

1. True
2. False
3. True
4. True

Worksheet 3
The Biblical Ice Age — Massive Storms and Snows

1. creationists
2. catastrophic
3. friction
4. precipitation
5. ice sheets
6. meteorological

Massive Melting

Discussion Questions

1. Glaciers in the Rockies melted at the end of the Ice Age, developing a massive lake blocked by an ice dam. The ice dam broke open and caused a flood across the Pacific Northwest.
2. Today we can view ripple marks or water lines thousands of feet high. We can also find boulders that were caught by the ice and carried hundreds of miles by the flood.
3. Erratic boulders were carried by large regional floods from one area to another. They consist of rock types that do not exist in the areas where they are located.

Worksheet 4
Another Ice Age

1. reoccurring
2. scientific basis
3. promise

Global Warming

1. active
2. melting
3. global warming
4. Little Ice Age
5. carbon dioxide
6. diatoms
7. ozone layer
8. stewards

Awesome Science Explore Dinosaur National Monument DVD Worksheet Answer Keys

Worksheet 1
Dinosaur National Monument

1. paleontologist
2. Allosaurus; Camarasaurus
3. sedimentary
4. Morrison Formation
5. petroglyphs
6. fossilized
7. evolutionary
8. Freemont
9. petroglyphs
10. Tower of Babel

Worksheet 2
Two Worldviews

1. naturalism; special creation
2. billions
3. evolved
4. Dinosaurs
5. Scientific laws
6. biblical
7. Creator; law-giver
8. corrupted
9. salvation

Worksheet 3
Finding Dinosaurs

1. sedimentary
2. quick burial
3. K-T Boundary
4. iridium
5. iridium
6. volcanoes

7. Flood

Worksheet 4
Dinosaurs at the Monument

1. evolutionists
2. different lizard
3. bipedal
4. predators
5. plants
6. Fall

Worksheet 5
What Happened to the Dinosaurs?

1. fossilized
2. violent
3. minerals
4. variety
5. rebellion
6. small horse
7. fossil record
8. habitat
9. simultaneous
10. quick burial
11. Marco Polo

Dinosaur Soft Tissue

1. True
2. True
3. False
4. False
5. True
6. True
7. False

Worksheet 6
Dinosaurs and Humans

1. Angkor, Cambodia
2. Peru
3. Turkmenistan
4. Congo
5. Job
6. behemoth
7. leviathan
8. dragons

Awesome Science Explore the Mammoth Site DVD Worksheet Answer Keys

Worksheet 1
About the Mammoth Site

1. Colombian mammoths; woolly mammoths
2. mammal fossils
3. underground springs; mineral water
4. sinkhole
5. flora fossil evidence
6. Secular; worldview
7. global Flood
8. geologic
9. Bering Strait
10. Columbian mammoths

Discussion Questions

1. The land would be better used for scientific study than for private use because of the fossils found there.
2. Young adult males are often kicked out of elephant herds, and they travel alone until they form their own herds. They would not have known the dangers of a sinkhole.
3. Christians use evidence from the Bible, specifically the Great Flood, which happened about 4,350 years ago.

Worksheet 2
Carbon Dating

1. Radiocarbon
2. Flood
3. sediments
4. half-life

5. assumptions
6. carbon 14
7. contamination
8. specimen
9. atmosphere
10. biosphere
11. Bible
12. magnetic field

Discussion Questions

1. The animals eat plants, which contain carbon 14 from photosynthesis.
2. A half-life is just a few thousand years, so if a million years passed, there would be no half-life left.
3. The secular assumptions leave room for too many errors, and they continue to make new discoveries that disprove their previous assumptions.

Worksheet 3
More about Mammoths

1. False
2. True
3. True
4. False
5. False
6. False

Discussion Questions

1. 100
2. Humans
3. Every 22 months
4. There were large grasslands near the coasts and they had few predators.

Worksheet 4

The Ice Age and the Bible

1. evaporation
2. volcanic
3. ice sheets
4. conditions
5. archeological
6. melted
7. fossil and historical record

Discussion Question

Secular scientists do not use the Bible as a real source or timeline.

Worksheet 5

Secular Ideas of Extinctions

1. Extinctions
2. humans
3. younger
4. theory
5. comet; asteroid
6. climate change

The Ice Age and the Bible

1. permafrost
2. climate
3. Temperatures
4. windstorms
5. silt
6. water table
7. Ice Age

Awesome Science Explore Arches and Natural Bridges
DVD Worksheet Answer Keys

Worksheet 1
About Arches National Park

1. Colorado
2. 2,000
3. 43
4. slickrock
5. 4,000
6. erosion
7. 10

Worksheet 2
History of Arches

1. Ice Age
2. Petroglyphs
3. Pictographs
4. Anasazi
5. tourist
6. monument
7. educational
8. 80,000

What Is an Arch?

1. True
2. True
3. False
4. False
5. True

Worksheet 3
What to See at Arches

1. Babel
2. Moab
3. Fiery
4. trails
5. peregrine

How did it happen?

1. 300
2. 150
3. 70,000
4. erosion
5. quick
6. days
7. catastrophic
8. 4,000

Worksheet 4
About Natural Bridges

1. White
2. erosion
3. Hopi
4. dinosaur
5. river

History of Natural Bridges

1. False
2. True
3. True
4. True
5. False

Worksheet 5
What to See at Natural Bridges

1. Sipapu
2. trails
3. Kachina
4. Hopi
5. Owachomo
6. Park

Worksheet 6
Geology of the Southwest and Natural Bridges

1. sandstone
2. sedimentary
3. seas

4. water
5. lower
6. turbulent
7. historical
8. catastrophic
9. surface
10. Flood
11. days
12. cavitation

Worksheet 7
Evidence of Dinosaurs

1. sauropod
2. 6,000
3. 6
4. Flood
5. stegosaur
6. 1,000
7. millions
8. varnish

Gospel message and closing

1. geologists
2. last
3. repent
4. short
5. timeline

Awesome Science Explore Mesa Verde and Chaco Ruins
DVD Worksheet Answer Keys

Worksheet 1
Mesa Verde & Chaco Ruins

1. thousand, global, Noah
2. created, race
3. one, Babel

Mesa Verde Introduction

1. canyons, mountain, plateaus
2. Colorado
3. green
4. Palace, largest
5. Thousand
6. Theodore Roosevelt
7. preserve

Discussion Questions

1. Answers may include golden eagles, coyotes, mountain lions, tarantulas, bullsnakes, rock squirrels, rabbits or any one of 200 species birds.
2. Early man's expansion across the landscape after the global Flood

Worksheet 2
What We Find at Mesa Verde

1. 600, 150
2. springs
3. largest
4. 2

Discussion Questions

1. Some have thought the people living here were short, between five to five and half feet tall. But the height of a doorway doesn't always mean the people were short or tall. Short doorways in northern Europe are even smaller, but they were that way not because of the size of people, but to keep the warm air in during the winter. If you compare them with European people of the same time period, they would have been about the same size.
2. high infant mortality rate
3. sandstone, mortar, wooden beams
4. local soil, water, ash

Worksheet 3
Who Built Them?

1. ancient, enemy
2. Asia, Ghengis Khan
3. 700, lived
4. drought
5. peaked, 850

Discussion Questions

1. corn, squash, and beans
2. The occupants lived in simple pit houses with a hearth, fire hole, and room for storage. They usually entered through the roof, by way of a ladder. The fact that the house was partly underground helped keep it cool in the summer and warm in the winter.
3. from the pottery found at these sites

Worksheet 4
Chaco Ruins

1. Chaco, social, Anasazi
2. Bonito, pretty, kivas
3. drought
4. aligned, primitive
5. Great
6. economic
7. Roads, Aztec, Salmon
8. lines, Peru
9. Pueblo, New Mexico

Discussion Questions

1. turquoise, parrots, macaws, copper bells, and other precious commodities
2. the lines are arrow straight, regardless of terrain. They go over mesas, up and down vertical cliff faces, and along ways that make them utterly impractical for use, by the casual or commercial traveler.
3. They did not worship the creator God, but instead, worshipped the creation itself.

Worksheet 5
Aztec Ruins

1. New Mexico, Aztec
2. Mesa, Chaco, commerce
3. decades, monumental
4. Verde, pottery
5. homes

Discussion Question

population pressure, climate changes, new enemies, disease, and shrinking natural resources

Worksheet 6
Man's Origins by Evolution

1. billion, nothing
2. Humans, Africa
3. civilization, Middle

Man's Origins by Evolution

1. Bible, God
2. garden, multiply, subdue
3. eternal, death
4. Nimrod

Discussion Questions

1. Satan rebelled against God and used a serpent to deceive Eve, and subsequently Adam, into sin.
2. likely due to genetic bottlenecks and mutational deterioration
3. God promised He would never flood the earth again and instituted the rainbow as a symbol of His promise.
4. Noah's sons-Shem, Ham, and Japheth
5. As the earth began to be populated, they started establishing a city and staying in one place at Babel and tried to defy God's command to be fruitful and fill the earth.

Worksheet 7
Tower of Babel

1. multiply, spread
2. one

3. Babel, rebelled, grace, destroying

Discussion Questions

1. To the east of Noah's farm they found the plain of Shinar and settled there, which is in modern-day Iraq, named for the old city of Erech, near Baghdad.
2. to make a name for themselves and not be scattered over the whole earth
3. 2350 B.C.
   2242 B.C.
   2234 B.C.
   2188 B.C.
   1921 B.C.
4. Aegyptus in Greek or Mizraim in Hebrew…its founder, who happens to be Noah's grandson!
5. The people spread out across the world into family and language groups.
6. The dispersion following the confusion of languages at the Tower of Babel.

Worksheet 8
A Question of Race

1. Darwin, favored, evolved
2. years, biologically, one
3. Races, exterminate
4. blood, Adam
5. variation, trivial
6. No, new, genetic
7. genes, characteristics
8. all, race, one

Discussion Questions

1. They were considered a type of missing link between apelike ancestors and the rest of mankind. They were hunted down, and killed because of evolutionary philosophy
2. There is only one race, the human race.
3. the amount of melanin in our skin
4. The Tower of Babel would have dispersed the family groups by different languages. They split apart, going their own way, finding different places to live. Because of the new language and geographic barriers, groups no longer freely mixed with other groups, and the result was a splitting of the gene pool. Different cultures formed, with certain features becoming predominant within each group simply because that is what their direct ancestors had coming out of Babel.

Worksheet 9
Establishing The Nations

1. earth
2. Adam
3. Shem, Abraham
4. Abraham, Noah, Adam
5. names, places
6. Chinese
7. Japheth
8. original, revisions

Discussion Questions

1. A genealogy is a list of fathers and their sons, through many generations.
2. Because many Christians and Christian leaders have accepted the secular dates for the origin of man and the universe, so they must work out ways that such dates can somehow be incorporated into the Bible's historical account.
3. "All Scripture is given by inspiration of God"
4. Japheth
5. ancient genealogies- ancient names and kingdoms

Worksheet 10
Division of Languages

1. 6900, 100
2. 4,200
3. animals
4. Secular
5. dividing

Discussion Questions

1. study of languages

2. All scripture is God breathed, meaning that God spoke through men, like Moses, to tell about the history of Man and what God did in the past.

Worksheet 11
Where Did Anasazi Come From?

1. Japheth's, Ham's, Shem's
2. European, native
3. Geographical
4. languages
5. families
6. Native
7. matching

Discussion Questions

1. a number of languages are all similar and have a common origin—ultimately dating back to Babel
2. This allowed for many changes over time as well as new words for new places, plants, and animals.
3. There is now evidence that the Clovis people were not the first on the continent.
4. All language/people groups dispersed from the Middle East. Some eventually made their way to Siberia, across a land bridge to the Americas, spreading throughout the continents.
5. the Americas and Australia

Worksheet 12
Where Did Anasazi Go?

1. south
2. Mexico
3. modern

Discussion Questions

All peoples of the ancient southwest, and Mexico, were of one language family.

Worksheet 13
Story of Man and Jesus?

1. Adam, Eve
2. Flood
3. languages
4. rebellion, forever
5. earth, sin
6. save, punishment
7. Sin
8. Repent
9. Ark, forever

Discussion Questions

1. what we find in the archeological record
2. what we find in genealogies, in cultures throughout the world.
3. what we find in archeology…a sudden explosion of culture and language
4. what we find in scripture, from the story of Noah's Flood and the Tower of Babel
5. death and eternal separation from Him
6. The wages of our sin is death, but the gift of God, is eternal life in Christ Jesus our Lord.

Worksheet 14
Closing Scene

1. thriving, mysteriously
2. dispersion, confusion
3. Bible
4. Jeremiah, turns, relent, disaster

Discussion Questions

1. He will bless them.
2. He will destroy it.
3. They worshiped the creation, rather than the Creator. Based on scripture, God would not have blessed this nation of ancient Americans. He would have sought to destroy and disperse them because of their rebellion. And this is exactly what we find about this culture.
4. to make sure our hearts and the heart of our nation are turned to God.

Awesome Science Explore Rocky Mountain National Park DVD Worksheet Answer Keys

Worksheet 1
Rocky Mountain National Park

1. North America
2. Thousand, British Columbia, New Mexico

Discussion Questions

1. In water and by water
2. The idea proposed by a creation geologist that the continents came to their current position in less than a year during the global Flood.

Rocky Mountain National Park History and Features

1. 1915
2. Woodrow Wilson
3. 415
4. 4,000, 12,000
5. 72

Worksheet 2
Mountain Building

1. Rockies, Alps, Himalayas, Appalachians
2. Black Hills of South Dakota
3. The Grand Tetons in Wyoming
4. Cascades, Aleutians, Hawaiian Islands

Discussion Questions

1. All of them
2. Several layers are pushed together on each end and begin to fold, causing them to rise and fall as the layers twist.
3. Pressures below the earth's surface push up layers and the layers get bent.
4. Layers shift down or up, like a knife went through the layers and cut them, so they shift vertically.
5. They form when molten lava comes up through the crust and pushes out onto the surface.
6. The Flood, in less than a year

Worksheet 3
Secular Ideas on the Rockies

1. Marine fossils
2. Metamorphic rocks, granites
3. Mount Everest
4. four millimeters
5. Pacific, North America, South American
6. The Pacific, the Rockies
7. Sedimentary rock, fossils
8. water

Discussion Questions

1. Sedimentary and volcanic rocks that have been changed by heat and pressure.
2. One plate sliding under another.

Worksheet 4
Biblical View on Rockies Formation

1. Yes
2. How much time it took
3. Tectonic plates began moving very quickly
4. The earth's plates were horizontally fixed
5. Catastrophic Plate Tectonics (CPT)
6. The earth was differentiated into core, mantle, and crust. The crust was separated horizontally into large stable blocks of rock above the waters, formed during the creation week. These types of rocks are mainly granites and metamorphic rocks. The ocean floor was primarily made up of mafic basalt, which is denser than granite.
7. Slabs of ocean floor broke loose and subducted along thousands of miles of pre-Flood continental margins. The pre-Flood supercontinent also broke up into fragments called plates.
8. Feet per second, which led to mantle convection which brought up massive amounts of heat from the interior of the earth
9. It affected the earth's core, which induced rapid reversals of the earth's magnetic field

10. Large plumes of molten rock that rose to erupt as vast volcanic outpourings
11. The rapid rising of basalt magma with steam came into contact with the ocean water and jettisoned ocean water with the steam high into the atmosphere where it condensed, creating intense global rain.
12. The new warm ocean floor rose quickly as the new less-dense magma that had produced it poured out from the mantle, causing the sea level to rise.
13. Fast moving ocean water transported sediment onto the continental margins. Tidal waves with ocean to land movement further transported sediments into the continental interior.

Worksheet 5
Fill in the Blank

1. Large earthquakes
2. Grand Canyon, northern Arizona, and Zion Canyon in Utah
3. subducted, continental edges
4. granites, metamorphic rocks
5. warmer
6. Ice Age

Discussion Questions

1. All of the cold, dense, pre-Flood oceanic floor was replaced with new, warm, less-dense, less-subtuctable oceanic crust
2. The continental margin stopped getting dragged down by the subducting slab, so it "bobbed" back up again.
3. They were sinking.
4. Receding flood waters caused channelized erosion.

Worksheet 6
Rate of Erosion

1. James Hutton
2. Bible
3. Erosion
4. mountains
5. 23
6. 9,000
7. Planation Surfaces
8. precipitation
9. precipitation, Ice Age, hundred
10. regional
11. larger
12. thousand
13. Bible

Discussion Questions

1. If current erosion rates are used, we should not find any old sediment in mountainous areas. Yet we find sediments from all ages preserved even on the highest peaks, such as at Mount Everest.
2. No, man's effect would need to have been several hundred times greater to make this argument valid.
3. During the flood the mountains rose and the ocean basins sank.
4. a giant sheet of water flowing quickly, spanning over many miles, erodes off entire sheets of sediments

Worksheet 7
The Great Sand Dunes

1. 750
2. unusual
3. mountain
4. climate
5. Ice Age
6. Silt
7. Hippos, giraffes, man, sand, desolate
8. hundred

Worksheet 8
Devils Tower

1. water, sediment, eroded
2. Wyoming

3. Power
4. 1906, Theodore Roosevelt
5. Igneous, 7, 4
6. 3,000
7. magma

Discussion Questions

1. 1. Devil's Tower is a large mass of volcanic rock, which intruded through sedimentary rock beds, but didn't actually reach the surface, producing a rounded bulge in the sedimentary layers above.
   2. Devil's Tower is a volcanic plug, or the neck of an extinct volcano.
2. They were formed quickly and solidified at a shallow level.
3. As the Rockies were being pushed up near the end stages of the Flood, large sheet erosion began planing off thousands of feet of sediments, exposing Devil's Tower.
4. Devil's Tower should have been leveled long ago

Worksheet 9
Conclusion

1. Flood, quickly
2. trusted, assumptions
3. Thousands
4. quick, Ice Age, Flood
5. Flood, major, judgment, Noah

Test Answer Keys

True North Grand Canyon —● Test Answer Key

Test 1

schist — a type of metamorphic rock that often contains reflective minerals, such as mica.

granite — a common and widely occurring type of igneous rock, commonly made up mostly of light-colored minerals, such as quartz and feldspar.

unconformity — a buried erosion surface separating two rock masses or layers of different ages, indicating that sediment deposition was not continuous; in the evolutionary model this also represents a lengthy interval of missing time.

land bridge — land connecting continents or landmasses that permits the migration of animals, often subject to temporary or permanent submergence.

Tapeats Sandstone — the lowest horizontal layer in the Grand Canyon, which sits directly on top of the basement rocks.

biological mutualism — a biological interaction between individuals of two different species, from which both individuals derive a benefit.

sheet erosion — the erosion of material by water flowing over land as a widespread mass instead of in definite channels or rills.

sapping — the process in which groundwater exits a bank or hillside laterally in the form of a seep or spring, eroding soil from the slope and often causing the collapse of material above.
riparian — the ecosystem of plants and animals living along a water source that are dependent on that water source.

ecosystem — A system that includes all living organisms (in an area as well as its physical environment) functioning together as a unit.

1. Great Flood
2. redwall limestone
3. the Anasazi
4. Many scientists believe the Anasazi left the area to find more favorable climate.
5. the San Francisco Peaks
6. 1908
7. time
8. A rapid flow of water in such huge quantities would have washed the rock debris away.
9. Nautiloids
10. the dipper or "water ouzel"

First Semester Test —● Test Answer Key

Test 2

1. Secular scientists say that the Colorado River carved the Grand Canyon over millions of years — yet the evidence does not support this idea. By tracing where the river begins and ends, and all the land it goes through, you can see that isn't possible. The river doesn't have the volume of water needed to do it even over a long period of time, there is no evidence the Kaibab Plateau rose at the time needed, and a river cannot run uphill — and certainly not the 4,000 feet needed. Most scientists now think that the Kaibab Plateau rose before the Canyon was cut — so something else must have been happening.
2. Erosion is a process by which water or another force can wear away the surface of something else — like water making a rock smoother.
3. Because of their worldviews. Secular science says the world took billions of years to appear as it does today because they say changes took place in a slow way over millions of years. Creation scientists say the world is very young based on the biblical record and the evidence they see in the world in the context of the biblical event of the Great Flood.

It shows that the history found in the Bible is true – and that the Bible is more than just a book of religious stories; it is a history book in and of itself. In many instances we see or learn something in the Bible, but it may have taken science hundreds of years to confirm it.

4. A worldwide Flood is rejected by those who do not believe the Bible is the ultimate source of truth and thus the scientific evidence is being filtered or interpreted through worldviews that do not include the Biblical record.
5. Biocrusts are often called "the glue of the desert" because they can absorb 10 times their volume in water and are especially important in the storage of rainfall. It is also a key factor in reducing both wind and water erosion.
6. cemented volcanic ash
7. Plant life above the ground may be destroyed in a fire, but many plants will come back after the fire.
8. Hayden Valley
9. A remnent of a vast, dry lake bed
10. Valleys are eroded, mountains rise, and lakes are formed
11. It is a fragment of space matter often made of iron and nickel.
12. Answers can vary. For example: Secular scientists think that it took millions of years, and processes like erosion from the Merced River and multiple ice ages over long periods of time.
13. Answers can vary. For example: The biblical model of one major Ice Age after the Great Flood supports the evidence we see in the world today versus many ice ages over time, an idea scientists have dozens of ideas on how they may or could have formed.
14. Answers can vary, but should focus on helping to determine the age of the Earth. For example: By noting the detailed genealogies from the Bible, you can form a timeline of the age of the Earth back to the point of creation, and they show an unbroken timeline of people – just like if you were tracing back your own family tree.
15. Answers will vary.
16. The earth trembles and the mountians smoke.
17. The evidence showed there could be massive changes in a short period of time.
18. Yes
19. They would have escaped a local flood.
20. The Flood currents would have brought diverse mammal fossils and layers.

True North Yellowstone — Test Answer Key

Test 3

caldera — a depression formed when a volcanic eruption empties the underground reservoir of magma, which is followed by the collapse of the roof of the magma chamber.

plateau — a landmass with a relatively flat summit area that is elevated above adjacent terrain and bounded on at least one side by a steep slope.

thermophiles — a "heat-loving" micro-organism that lives in or around hot springs and hot pools that include bacteria, viruses, archaea, and eukarya.

supervolcano — a very large volcano capable of producing an eruption of more than 240 cubic miles of material.

erosion — the displacement of solids (soil, mud, rock, and other particles) by the action of wind, water, or ice.

biodiversity — the variety of life within an ecosystem or community.

fault scarp — a topographic surface expression of faulting attributed to the displacement of the land surface by movement along a fault.

contact — the boundary surface between two types of ages or rocks; (contacts are either "conformable" if they represent no gap of time between them, or an "unconformable" if they are thought to represent a gap of time or erosion between the rocks).

1. 8,000 feet
2. 10,000
3. God's word, the Bible
4. One and a half
5. Shortly after the receding of the Flood
6. Thermophiles

7. Granite and gneiss
8. Glaciers
9. They are all part of the water cycle.
10. 70 percent

Second Semester Test — Test Answer Key

Test 4

1. The Grand Tetons
2. Mud pots, fumaroles, hot springs, geysers
3. Glaciers in the Rockies melted at the end of the Ice Age, developing a massive lake blocked by an ice dam. The ice dam broke open and caused a flood across the Pacific Northwest.
4. Today we can view ripple marks or water lines thousands of feet high. We can also find boulders that were caught by the ice and carried hundreds of miles by the flood.
5. Erratic boulders were carried by large regional floods from one area to another. They consist of rock types that do not exist in the areas where they are located.
6. Answers will vary.
7. Sedimentary
8. The dust from an asteroid does not spread very far, and the amount found in this layer in various places around the earth is just too much for a single impact from an asteroid fragment to produce.
9. The land would be better used for scientific study than for private use because of the fossils found there.
10. The animals eat plants, which contain carbon 14 from photosynthesis.
11. A half-life is just a few thousand years, so if a million years passed, there would be no half-life left.
12. Answers may include golden eagles, coyotes, mountain lions, tarantulas, bullsnakes, rock squirrels, rabbits, or any one of 200 species birds.
13. Early man's expansion across the landscape after the global Flood.
14. Some have thought the people living here were short, between five to five and half feet tall. But the height of a doorway doesn't always mean the people were short or tall. Short doorways in northern Europe are even smaller, but they were that way not because of the size of people, but to keep the warm air in during the winter. If you compare them with European people of the same time period, they would have been about the same size.
15. The idea proposed by a creation geologist that the continents came to their current position in less than a year during the global Flood.
16. Several layers are pushed together on each end and begin to fold, causing them to rise and fall as the layers twist.
17. They form when molten lava comes up through the crust and pushes out onto the surface.
18. Receding flood waters caused channelized erosion.
19. A giant sheet of water flowing quickly, spanning over many miles, erodes off entire sheets of sediments
20. They were sinking.

Map of State Parks and Monuments in the Western United States

Map Answer Key

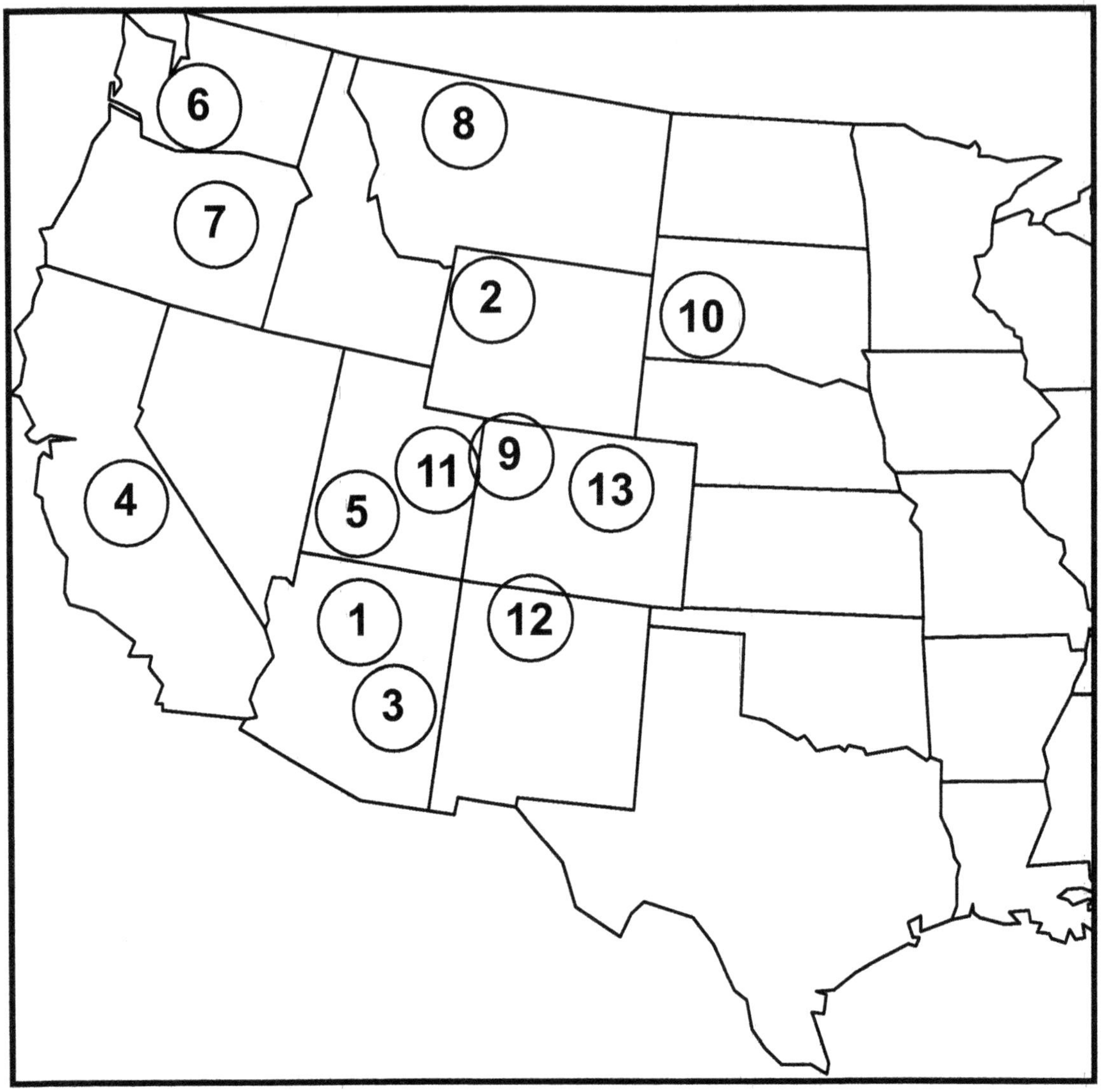

1. Grand Canyon National Park
2. Yellowstone National Park
3. Meteor Crater and Petrified Forest National Parks
4. Yosemite National Park
5. Zion National Park
6. Mount St. Helen's
7. John Day Fossil Beds National Monument
8. Glacier National Park
9. Dinosaur National Monument
10. Mammoth Site
11. Arches National Park and Natural Bridges National Monument
12. Mesa Verde and Chaco Ruins
13. Rocky Mountian National Park